E **prise**

for Life Scientists

Dedication

To Louise, Margaret and Ellen,
and Sue, Katy and Helen.

Enterprise

for Life Scientists

Developing Innovation and Entrepreneurship in the Biosciences

Edited by

David J. Adams

Faculty of Biological Sciences and Higher Education Academy Centre for Bioscience, University of Leeds

and

John C. Sparrow

Department of Biology, University of York

Scion

© Scion Publishing Ltd, 2008
First edition published 2008

ISBN 978 1 904842 36 1

A CIP catalogue record for this book is available from the British Library.

Scion Publishing Limited
Bloxham Mill, Barford Road, Bloxham, Oxfordshire OX15 4FF
www.scionpublishing.com

Important Note from the Publisher
The information contained within this book was obtained by Scion Publishing Limited from sources believed by us to be reliable. However, while every effort has been made to ensure its accuracy, no responsibility for loss or injury whatsoever occasioned to any person acting or refraining from action as a result of information contained herein can be accepted by the authors or publishers.

Typeset by Phoenix Photosetting, Chatham, Kent, UK
Printed by Gutenberg Press Ltd, Malta

Contents

Chapter 3 – Protecting Ideas   52

Louise Byass

Chapter 4 – Researching Ideas   81

David Wilkinson & Amanda Selvaratnam

Chapter 5 – Communicating Ideas   103

Samantha Aspinall & David Wilkinson

Chapter 6 – Starting up a Business   127

Alison Price & Ted Sarmiento

Chapter 8 – Funding your Ideas 181

David Baynes & Louise Pymer

Chapter 9 – Regulation in the Biosciences

201

Nick Medcalf & Bob Pietrowski

Chapter 10 – Ethical Issues

223

Rob Lawlor

Foreword

In recent years universities have invested in creating a culture which recognises and values the importance of innovation, enterprise and effective knowledge transfer. However, enterprise is not a universally recognised academic subject and is only slowly entering the curriculum of many universities. Yet acquiring and developing enterprise and entrepreneurship skills and competencies are fundamental to successfully managing personal, business and work opportunities in order to make an impact in the future as an enterprising employee and/or successful business owner.

Enterprise education has two important facets; it seeks to develop a broad range of entrepreneurial skills, attitudes and behaviours such as innovation, creativity, risk-management and risk-taking and a can-do attitude, while it has to deliver on developing knowledge about enterprise and entrepreneurship itself.

The authors have provided key material that will help students develop their ability to generate, recognise and seize opportunities. This book represents a novel and unique approach to the teaching of enterprise skills to life scientists.

Enterprise for Life Scientists will become a required text for undergraduate courses. Additionally, there is also real value here for the practitioner as it contains material essential to postgraduates, post-docs and academics (from both educational and industrial settings) who wish to pursue the successful commercial exploitation of their ideas.

Dr Julian White
Chief Executive, White Rose University Consortium

Preface

Bioscience students today must acquire a comprehensive knowledge and skills base. In addition, they are expected to develop an understanding of how their science can be exploited in a commercial setting. In the UK, this expectation is driven from three sides: a government that wants to exploit and retain the ideas developed by the bioscience community; industry which needs graduates well-versed in commercial skills; and fee-paying students wanting the best possible preparation for their future careers.

These three groups are all interested in biotechnology, as the industrial applications of bioscience are generally known, and this sector is currently estimated to contribute £200–300 billion to the global economy each year. The sector continues to grow and this expansion means that biotechnology has become an increasingly important part of the world economy that has created more and better opportunities for bioscientists to use their discipline for the public good.

Enterprise for Life Scientists is the first textbook to seek to equip students with the skills necessary to take part in the industrial application and commercial exploitation of bioscience. These are not just the skills needed by budding entrepreneurs; many are transferable to any working environment. For example, the book will help students develop their creative potential and learn how to communicate and network effectively. To highlight 'real world' applications of the key skills described, the book also contains case studies and biographies of successful entrepreneurs.

The authors include bioscience professionals, who have a variety of applied science and commercial roles, and experienced university teachers. Together they explain the increasing importance of knowledge transfer and provide an introduction to key regulatory and ethical issues in the biosciences. They guide the reader through the generation, protection, funding and communication of ideas. The role of the business plan is described and advice provided about how to start up a business. The book therefore provides a comprehensive account of how to develop and exploit ideas prior to embarking on the more generic activities of marketing and selling the finished product.

This is not just a text for undergraduates, as the depth of coverage in many of the chapters makes the book a valuable reference source for lecturers and postgraduate students grappling to come to terms with the commercialisation of their research. However, it is today's bioscience students, embarking on the growing number of university courses incorporating elements of 'bioenterprise', who will benefit most as they develop the skills required to meet the tremendous challenges and opportunities that will surely come their way in the 21st century.

David J. Adams and John Sparrow
August 2007

About this book

The book is structured to provide an initial introduction and explanation of knowledge transfer issues. A series of chapters then take the reader through a range of steps between idea generation and exploitation. Finally the reader is made aware of issues involved in starting up a business and regulatory and ethical matters that may have to be taken into account along the way.

If you are an undergraduate or postgraduate student of the biological sciences you will find this book an invaluable and unique introduction to enterprise in the curriculum. You will learn how to generate, protect, research, communicate and fund ideas, and you will be informed of key issues in bioenterprise including knowledge transfer, important regulatory concerns and relevant ethical considerations. Similarly, postdoctoral fellows will find the chapters on idea generation and exploitation of real value and will benefit greatly from an enhanced understanding of knowledge transfer issues in a university setting. Finally, academics wishing to develop enterprise learning in the curriculum or to commercialise ideas that stem from research laboratories will be able to make good use of advice relating to knowledge transfer, IP protection and the complexities of starting up and maintaining a spin-out company.

Contributors

David J. Adams

Faculty of Biological Sciences and Higher Education Academy Centre for Bioscience, University of Leeds, Leeds LS2 9JT

Kathy Armour

Enterprise and Innovation Office, University of York, York Science Park, York YO10 5DG

Samantha Aspinall

Faculty of Biological Sciences/Enterprise Centre, University of Leeds, Leeds LS2 9JT

David Baynes

Biofusion plc, The Sheffield Bioincubator, 40 Leavygreave Road, Sheffield S3 7RD

Shane Booth

ANGLE plc, Mansfield i-Centre, Hamilton Way, Mansfield NG18 5BR

Louise Byass

IP Exploitation Manager, Central Science Laboratory, Sand Hutton, York YO41 1LZ

Andrew Ferguson

Enterprise and Innovation Office, External Relations Department, University of York, Heslington, York YO10 5DD

Paul Grimshaw

Faculty of Biological Sciences, University of Leeds, Leeds LS2 9JT

Rob Lawlor

Inter-Disciplinary Ethics Applied, A Centre for Excellence in Teaching and Learning, University of Leeds, Leeds LS2 9JT

Nick Medcalf

Bioprocessing Manager, Smith & Nephew Research Centre, York Science Park, Heslington, York YO10 5DF

Robert Pietrowski
David Begg Associates, The Georgian House, 22/24 West End, Kirkbymoorside, York YO62 6AF

Alison Price
CETL Director, The Institute for Enterprise, Leeds Metropolitan University, The Grange Coach House, Headingley Campus, Leeds LS6 3QS

Louise Pymer
Biofusion plc, The Sheffield Bioincubator, 40 Leavygreave Road, Sheffield S3 7RD

Ted Sarmiento
The Institute for Enterprise, Leeds Metropolitan University, The Grange Coach House, Headingley Campus, Leeds LS6 3QS

Amanda Selvaratnam
Continuing Professional Development Manager, Innovation Centre, York Science Park, York YO10 5DG

John Sparrow
Department of Biology, University of York, Heslington, York YO10 5DD

David Wilkinson
Leeds University Business School, Clarendon Road, Leeds LS2 9JT

Abbreviations

AIM	Alternative Investment Market
BBSRC	Biotechnology and Biological Sciences Research Council
BIA	Bioindustries Association
BIO	Biotechnology Industry Organisation
BSF	Biosciences Federation
CDA	confidential disclosure agreement
CI	competitive intelligence
DEFRA	Department for Environment, Food and Rural Affairs
EC	European Community
EPC	European Patent Convention
FDA	Food and Drug Administration
FOE	Friends of the Earth
GMO	genetically modified organism
GMM	genetically modified micro-organism
GMP	good manufacturing practice
HSE	Health and Safety Executive
IP	intellectual property
IPR	intellectual property rights
JETRO	Japan External Trade Organisation
LSE	London Stock Exchange
MTA	material transfer agreement
NDA	non-disclosure agreement
PCT	Patent Cooperation Treaty
PEST	political, economic, social, technological
R&D	research and development
QMS	quality management system
SME	small and medium-sized enterprises
SWOT	strengths, weaknesses, opportunities, threats
USP	unique selling point

Chapter 1
Knowledge and Technology Transfer

Shane Booth & Kathy Armour

1.1 – The bioscience business

Around 30 years ago there began a revolution in the life sciences with a series of major breakthroughs in recombinant DNA, monoclonal antibodies and other technologies. The new tools that became available gave bioscience researchers a power and scope utterly undreamt of a generation before and led them into new activities, fields and cultures. Biology has long had practical applications in areas such as medicine and agriculture but, with the new developments, these have recently exploded in importance leading to a re-positioning of bioscience as a major force in commerce and economies.

The expansion of commercial bioscience is important partly because of the benefits it generates in the form of new products and services (these ranging from revolutionary new drugs to disease-resistant crops), and partly because of its scale. Biotechnology, as the industrial applications are generally known, is now maturing as a highly productive global industry. Depending upon precisely how it is defined (i.e. the range of activities, products and services that are considered) the biotechnology sector is currently estimated to contribute US$100–150 billion to the global economy each year in money earned from sales and investment. To put this into perspective, this performance is of a magnitude greater than that of some entire (albeit small) economically successful countries. For example, the entire economy of Scotland (i.e. the amount of money being earned and spent by and within the country) is equivalent to about US$100 billion. While biotechnology is young, it is growing rapidly. In 2005, the public companies (i.e. those mature and successful enough to trade their shares on stock exchanges) operating in this sector across the world generated collective revenues (i.e. proceeds from sales) of US$63.1 billion, an increase of 18% on the previous year[1].

This massive expansion in biotechnology has been part of a wider trend for new technologies to generate greater prosperity. In order to explain how academic sciences have become such powerful economic drivers it is useful to go back to when all of this started; a time when things were far less encouraging.

1.1.1 Industry old and new

For those who grew up in them, the 1970s and 1980s were often worrying times. Both decades saw the UK's industry suffer terrible problems with strikes, problems with access to energy, and huge competition from overseas. The closures of once proud firms, unemployment, social devastation, and the political unrest these caused, generated a widespread belief that the country was in a decline from which it would never recover. A generation grew up far from sure that prosperity and economic security would ever return.

But the UK recovered. It is once again a fiscally secure nation, its economy being the fourth largest in the world and second largest in the European Union, with unprecedented opportunity and quality of life for its citizens. This recovery was partly the result of reform of the social, economic and legal landscape by successive governments during the 1980s and 1990s, but also because there emerged new industries to take the place of the traditional ones that were in decline. Much of this new industry had its origins in revolutionary developments in science and technology. To understand how this happened we need to look at the nature of markets.

1.1.2   Markets and technology

It is not uncommon for experience with aggressive sales behaviour to lead people to believe that commerce is about making people do things they don't want to do. Though it would be wrong to deny that this goes on, the fact is that successful industry is based on quite the opposite. There has always been demand for solutions to all kinds of human need. Commerce comes from the need or aspiration of people and organisations to do or achieve things they can't do without some kind of tool or assistance: things that bridge gaps between aspiration and capability often have the potential to become successful products and services, particularly where the need for them is extensive.

If we accept then that successful commerce is rooted in effective solutions to human need, it follows that solutions to particularly important or widespread problems can have consequently high potential as things that can be sold. Technology, by its very nature, can provide such strong solutions, and in bridging important capability gaps in this way can have high commercial impact. Huge commercial success can arise when technology:

- solves problems that are both difficult and important, for example, the drug Prozac's success in treating mental health
- makes possible things that are so useful or desirable that demand for them is extreme, for example, domestic computer technology that has enabled the development of the advanced gaming industry

Figure 1.1: Capability gaps and markets.
This illustrates the simple but crucial fact that commercial success comes when a product or service successfully bridges a gap between an aspiration and capability. Such bridges often come in the form of solutions provided by new technologies. ▶

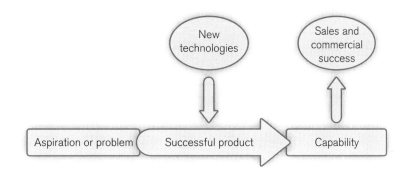

So technology is very important as a basis of business (see *Figure 1.1*) and *Table 1.1* shows ways in which technology is effective at levering new business opportunity.

Table 1.1: Ways in which technology generates new business	
Business development mechanism	Examples
New approaches to existing markets	Development of recombinant human insulin. Insulin for use by diabetics had for decades been sourced from animal tissue. Efficiencies in its production and the reduction in side effects because of its much closer match to the native form made human insulin prepared in culture from isolated human genes a very effective product, allowing its creator Genentech Inc. (www.gene.com) access to this huge market.
Expanding existing markets	The commercial availability of restriction enzymes and related tools, starting in the late 1970s, enabled many more research groups to use gene splicing techniques than had been possible before. The market for specialised consumables for molecular and cellular biology research consequently expanded rapidly, with major benefits for pioneers such as Bethesda Research Laboratories (now Invitrogen Corporation (www.invitrogen.com)) and Stratagene Corporation (www.stratagene.com). The advent of oral contraception in the 1960s, in addition to causing major social change, was a way for pharmaceuticals manufacturers to expand their markets beyond people who were sick.
Creating new markets	Personal computers and the internet. The benefits of internet access have created markets for personal computers across age and social groups that previously had little interest in them.

The nature of contemporary society drives continual new opportunity through constant evolution of new markets. Increasing western life expectancy, lifestyle expectation, globalisation, intense competition from countries with low labour costs, and changing consumer tastes, all contribute to a rapid pace of change. Therefore new market opportunities are constantly being created. The strength of the products and services that are most successful in addressing these modern markets commonly comes from their having:

- unique selling points
- competitive advantage
- high added value – special features and capabilities additional to their core characteristics that make them particularly attractive

Creative thought and ideas are needed to conceive such new products and processes, and leading edge business is therefore intrinsically linked to *innovation*. For the purposes of this chapter, innovation can be considered to be simply the practical translation of ideas into new or improved products, services and processes, but innovation is further defined and discussed in *Chapter 2*.

1.2 — Academia as a business generator

1.2.1 Technology and knowledge transfer

It is inescapable that most leading edge technology has its origins in universities, government applied research centres, and teaching hospitals. Academic research has therefore emerged as a generator of new business development potential and, during the last two decades or so, research institutions across the developed world have become important business organisations in addition to their traditional activities. This is far from a peripheral activity: academic institutions are now recognised as serious business concerns, and on a university campus one is now as likely to see a commercial manager in a suit as a researcher in a white coat.

The commercial potential of academic capability resides in both expertise and technology, and the process in which university technical capability is applied and/or commercialised has come to be called knowledge or technology transfer. The term generally refers to processes by which knowledge, expertise and skilled people transfer from academic to commercial environments, and contribute to business development, economic competitiveness, effectiveness of public services and policy, and quality of life. A measure of the importance with which academia-based business is now regarded, is that the term is now used as a label for what is regarded as a whole area of commerce: the knowledge transfer sector.

1.2.2 Academic and industrial roles

How does academia do industrial work? After all, it is obvious that the activities that occupy university staff and the personal characteristics that contribute to their ability in teaching and research are different to the nature of industry and business management. The answer, perhaps obviously, is through linkage and combination of academic skills with those used in management of business and facilities in industry.

Sometimes technology is developed specifically for an applied need; alternatively its commercial potential may be fortuitous and identified retrospectively. Either way, technology rarely comes in a form ready for use in professional settings or for sale into commercial markets. It needs to be developed into a market-ready form and conveyed to its markets through a proper commercial process. *Figure 1.2* illustrates the range of different

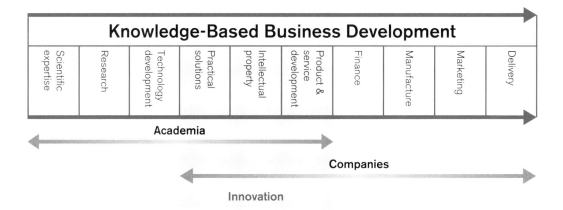

Knowledge-Based Business Development

Scientific expertise | Research | Technology development | Practical solutions | Intellectual property | Product & service development | Finance | Manufacture | Marketing | Delivery

Academia

Companies

Innovation

Figure 1.2: The knowledge business development chain.
Research and industrial organisations each have important roles to play in the initiation of knowledge or technology-based products and services and their eventual manufacture and delivery. There is an important cross-over area in which the two sectors interact most closely, and where the crucial process of innovation tends to take place. ▲

activities, skills and resources required in order for this to happen, and how neither the academic nor industrial sectors can provide all these capabilities independently.

The need for this extended range of capabilities, some of them highly specialised, means that this field of business usually involves partnerships between academic organisations and existing or newly created companies.

Figure 1.2 also highlights an area of overlap in the roles provided by academic and industrial partners. The creative interaction between the two sectors occurs predominantly in the generation of solutions to problems, consequent intellectual property, and translation into new products and services. These three processes represent innovation. Note that though it is tempting to consider academia (based as it is on intellectual activity) as being the source of innovation in knowledge business development, this is only partly the case. The innovation takes place at the interface between academia and industry, precisely where there is overlap in their roles.

In general, the *direction* of knowledge transfer is *from* the academic *to* the industrial sector. But the *process* is two-way: expertise is deployed from both the industry and academic bases. Remember, for example, that research scientists aren't restricted to academia; industry employs them too. Synergy arises when knowledge flows between scientists in the two sectors because of the different perspectives, technical experience and scientific knowledge they bring to problems. Synergy is also generated from convergence of the different core activities that exist between the two sectors. Creating new ideas is all in a day's work for academic researchers, whereas identification of market opportunity, and intellectual rigour in assessing the true market potential of new technologies are strengths of personnel with business backgrounds. The bi-directional interaction is also reflected in the identification of new commercialisation opportunities. This can be driven from academic research when new technical advances generate prospects for product and service development (technology push), and when commercial openings with possible technological solutions are recognised by the private sector (market pull).

1.3 – Knowledge transfer mechanisms

The ways in which knowledge transfer partnerships between academia and industry take place fall into two types:

- the creation and/or deployment of technical skills and expertise
- the creation of intellectual property

These two activities differ in that the former provides resources and operational capability for industry whereas the latter provides the actual entity that can be commercialised, i.e. that can be the basis of a new product or service.

1.3.1 Knowledge transfer of academic skills and expertise to industry

Provision of skills and expertise to industrial need are very close to university core activities of teaching and research, and include:

- generating graduates and postgraduates with technical skills of value to commercial organisations in technical and other roles
- student placements and secondment of academic staff to industry
- technical consulting in which academic staff and departments are contracted by industry to provide technical advice and to solve specific technical problems
- collaborative research in which academic and industrial research groups combine their expertise and facilities to solve technical problems and develop products and services

Case study 1.1 shows an example of collaborative research where an academic–industrial partnership successfully solved an important need for the company involved, and led to an extension of its commercial reach and business development capability.

1.3.2 Knowledge transfer as intellectual property

The nature of intellectual property (IP) is described in detail in *Chapter 3*. At this stage it is sufficient to acknowledge that it is more than just technology. Intellectual property arises where a product of human intellect is unique, non-obvious and has value in a marketplace, and so can include inventions, ideas, know-how and technical information, documents, music, and a variety of other things. Its exploitation in a marketplace is usually carried out through one of the following commercial mechanisms:

Case study 1.1: Authentix
A highly successful collaborative commercial R&D project

Adulteration of fuel is a growing issue. Illicit trading of fuel is a global problem costing international governments and industry billions of pounds annually in lost tax revenues and sales. Whilst current technology allows forensic fuel testing to be conducted in the field, a rapid forensic test would provide commercial benefits, including the reduction of costs and the improvement in chain of custody evidence. Such an advance was achieved when a collaborative R&D project was established between Authentix Ltd, an international authentication company with a UK R&D base, and the Analytical Science research group of the Department of Chemistry, University of York. The alliance was so successful that it won the Best Knowledge Transfer Partnership (KTP) Programme Award, sponsored by the Department of Trade and Industry for the 2005 competition.

The project aimed to develop a novel method for detecting levels of security markers in fuels using a rapid, automated and reliable test that could be carried out in the field. Academic and industry collaborators, each bringing a unique blend of expertise to solve the problem, recruited a highly motivated graduate to work on the agreed project. A collaborative agreement between the university and the company established the ownership of intellectual property created in the course of the project.

A prototype test system for detection of markers in fuels was developed that was based on microfabrication technology and solid state optical detection. Such technology allowed a covert marker to be detected and quantified forensically within seconds by non-technical operators. The absence of the need for manual handling improved the robustness of the system, offering significant performance benefits. An additional benefit was that the microscale device resulted in low levels of waste. Success with this method led to a patent being filed and the award of a follow-on project to develop a commercial version of the detector.

Not only were advances made in the core opto-electronics project but a novel reader was developed for use in security inks which emerged from the advances in the main project. Two contracts with international pharmaceutical companies worth over £1.5 million per annum and sales to the US government ensued.

The innovative prototype microfluidic fuel testing device has positioned Authentix and the University of York at the forefront of microseparations and nanotechnology. The mix of academic and industry-based technological expertise and lateral thinking contributed to the success of this collaboration. Knowledge transferred from the university base to the company in the fields of microfluidic science optics and electronics has enhanced the position of the company in these technical areas. On the other hand, the university contributors have also benefited from the initiation of new research and industrial collaborative programmes and teaching materials that allow direct visualisation of processes to complement theory, and the demonstration of how pure science can be used to solve real-life problems. In addition, the academics gained commercial experience in order to set up a university spin-out company in collaboration with colleagues involved in instrumentation development.

This case study shows that opportunities can be sought that require the application of academic rigour and experience to commercial and technological challenges. It highlights the use of technology as a mechanism of market opportunity, the need for market focus, and the opportunities for high growth and economic development that technology can bring. Essential to the successful mix was the recognition of the project risks and practical problems, the seeking of appropriate advice and expertise, and the blend of people with the appropriate attitude working within an agreed project framework.

- sale of the IP to a commercial organisation in return for a direct payment
- licensing of some or all of the rights to exploitation of the IP to a company in return for payments that are commonly linked to sales or profits from the product or service – remuneration is commonly monetary payments, but may also include shares in the company
- creation of a new company (often called a 'spin-out') to develop the product or service and deliver it to its markets

See *Chapter 6* for a more detailed discussion of the creation and nature of technology-based start-up companies.

Depending on the specific situation, research organisations may negotiate commercial terms with existing companies, which can range from small specialised firms to major corporations operating on a global scale. Alternatively, they may create a completely new company. (See *Chapter 6* for further discussion, including on the generation of *alliances* between different types of partner.)

Note also how the process is based on IP and, crucially, the rights to its commercial exploitation (intellectual property rights (IPR)). Creation of IP is one thing, but making money from it requires securing ownership and control of the exploitation rights. Consequently, a big part of knowledge transfer is the protection of IP in the form of patents, copyright, design rights, trademarks, etc. (see *Chapter 3*).

1.4 – Commercialisation of bioscience

Bioscience (and here the terms biology, bioscience, life science and biotechnology are used more or less interchangeably), long a successful discipline, has seen a revolution in its progress since the development of recombinant nucleic acid techniques in the mid to late 1970s. Together with some other major technical advances (such as monoclonal antibody technology), gene splicing seeded a wide portfolio of new analytical and preparative techniques of tremendous power. The central argument that commercial success is based on the bridging of capability gaps, helps to explain the fact that this scientific revolution has generated a consequent commercial one.

Earlier in the chapter it was explained how particularly difficult capability gaps often require *technical* solutions, and how their importance can generate a lot of demand for them. One of the best examples of this is human disease. The very nature of diseases means that they are understood only through a scientific approach, and their treatment has been made possible largely on the basis of science-based techniques. Diseases generate a lot of demand: people who are ill or injured have a level of desire for successful treatment that often exceeds any other aspiration. Consequently, the commercial sources of treatment, the pharmaceutical, medical device and hospital supplies industries, have become extremely successful. Modern

Table 1.2: Bioscience-derived products solving difficult clinical problems

Product	Background	Company
Epogen – recombinant erythropoietin	Anaemia is a debilitating disease caused by poor oxygen transport because of low numbers of red blood cells. Characterised by severe fatigue, it is a serious problem in renal patients receiving dialysis. Erythropoietin is a 165 amino acid glycoprotein, produced in kidneys, that stimulates red blood cell production through division and differentiation of erythroid progenitors in bone marrow. Cloning of the gene for this protein and subsequent translation in cultured mammalian cells has enabled it to be produced on an industrial scale. Marketed as Epogen, it was among the first biotechnology-derived drugs to reach the markets and is arguably the most successful, having achieved sales of well over US$20 billion since its release in 1989.	Amgen Inc. (www.amgen.com)
Rituxan – the first monoclonal antibody drug for cancer	A long-standing problem in cancer treatment is that treatments such as radiotherapy and chemotherapy are often insufficiently selective, damaging healthy cells and tissues as well as cancerous cells. Monoclonal antibodies, by their very nature, are exquisitely selective, binding exclusively to their target molecules. Non-Hodgkin's lymphoma involves malignancy of B cells carrying a cell surface protein called CD20. The Rituxan antibody is specific for the CD20 antigen and therefore binds only to the B cells that bear it, causing the immune system to attack and eliminate them, and only them. Healthy CD20-bearing B cells are also killed, but are replaced by the natural cell development process from bone marrow stem cells. Sales of Rituxan have exceeded US$8 billion since its US launch in 1997.	Genentech Inc. (www.gene.com)
Vitravene (fomivirsen) – the first antisense drug	Antisense techniques involve the production of nucleotide sequences that are complementary to specific mRNAs. Being complementary, they bind to the mRNA and can inhibit its trans-lation into protein. The technique therefore potentially enables the expression of any gene of interest to be inhibited – essentially switched off. Although technical problems cause the technique to be unreliable in some situations, it is a powerful research tool and has useful applications as a drug. HIV/AIDs patients are highly prone to secondary infections and other problems because of the severely depleted state of their immune systems. The drug Vitravene is composed of antisense nucleotide sequences that inhibit expression of two of the virus's genes. Although its sales have been restricted because of the success of general treatment of HIV infection, Vitravene proved that antisense technology can be successfully deployed as a clinical tool and showed how bioscience can bring completely new approaches to difficult problems.	Isis Pharmaceuticals Inc. (www.isispharm.com)

bioscience's technical power has added to this market-driven potential, enabling major growth through dramatic new product development ability. *Table 1.2* contains examples of how the bioscience sector has created technical solutions to challenging medical problems.

The term biotechnology has come to be closely associated with the pharmaceuticals industry and new drug development, but an important aspect of the field's commercial potential is that it also has far wider applications. As genetic engineering techniques began to emerge in the mid-1970s, much attention focused on practical applications for these procedures and their potential as manufacturing tools. Much of genetic engineering is about identifying, moving and expressing genes, and one of the most obvious opportunities was the production of useful proteins on an industrial scale, through the expression of genes in microbial or eukaryotic cells. This opportunity has been fulfilled, but many other practical approaches have also been developed and this is reflected in the breadth of applications and industrial sectors that now involve bioscience. Examples of important applications aside from new drug development include the following:

- *Instruments and reagents for medical and veterinary diagnosis.* Identification of molecules whose presence, position or changing characteristics are highly indicative of disease and prognosis, and the availability of recombinant proteins and monoclonal antibodies that enable these to be practically exploited, have all advanced this field.
- *Crime detection.* Genetic fingerprinting has caused a revolution in the forensic identification of perpetrators of crimes and the elimination of innocent people under suspicion.
- *Agriculture.* Recombinant techniques can be particularly powerful when deployed in plants. Examples of recombinant applications include the use of antisense techniques to endow disease resistance, control of ripening, and the ability of crops to withstand herbicides that control weed infestation.

Bioscience is a major part of university commercialisation, largely because of how active and technically strong life science departments and medical schools have become as a result of the bioscience revolution. Academic bioscience constantly generates new technology representing practical applications and solutions to industrial problems. Some applications may have obvious commercial potential on such a scale that they quickly attract the attention of large companies. Where this is the case, and where the technology can be developed into a new product quite quickly, the institution may elect to commercialise the IP through a license agreement with the company expressing the interest. For further discussion of licence agreements, see *Chapters 3* and *6. Case study 1.2* describes the highly successful example of licensing the technology for DNA fingerprinting.

Case study 1.2: Cellmark Diagnostics
A classic example of licensing of university intellectual property

The emergence of biotechnology promised revolutionary developments from its practical applications, and one area in which this has genuinely proven to be the case is the field of DNA fingerprinting.

Restriction enzymes break genomic DNA into very large numbers of fragments of hugely varying lengths. Smaller fractions of these can be identified and (crucially) visualised by hybridisation with labelled probes made from certain individual isolated fragments. The lengths of certain DNA fragments revealed in this way frequently vary amongst individuals, reflecting genetic variation (for a full technical description see Brown[2]). In the mid-1980s Alec Jeffries (now Professor Sir Alec Jeffries) at the University of Leicester was using this technique as a tool to research human genetics, and in a striking intellectual leap he recognised that a seemingly incomprehensible level of variation came from a sensitivity so exquisite that it absolutely matched individual identity. It was immediately obvious that the technique had applications in crime detection, immigration and paternity testing. This potential was proven when it successfully resolved several difficult and important criminal and other identity cases. For several years, the Jeffries research group provided the technique as a contract service, but the importance of the growing portfolio of applications and the work load eventually made it necessary to find alternative approaches to its provision and expansion.

The technique was licensed by the University of Leicester to ICI (now AstraZeneca) who further researched the scientific background to the technique, its range of applications, and quality assurance methods crucial to its capability of formal, absolute proof of identity and/or presence at a place or event. In 1987 this new part of ICI's business became Cellmark Diagnostics. Cellmark has been extremely successful in continuous development of the technique, and in providing it as a forensic service in crime, paternity, and more recently in agricultural applications. Now known as Orchid Cellmark, the Company operates from four locations in the US and one in the UK, and generates global sales of in excess of $60 million per annum.

The case of Cellmark's history and development is an excellent example of university research providing a solution to a range of hugely important practical problems, and creating an entire new commercial market in doing so. In addition, it illustrates how the transfer and translation of the technology into a commercial operation requires the combination of technical and commercial management skills, and how the operational challenge of a highly specialised new field was solved by creation of a new company that developed the management and processes to match new operational and market needs.

The licensing approach has the advantage of being comparably low in risk to a university because further development of the project, and its costs, are borne by the partner company. Passing on this proportion of the cost and consequent risk means, however, that there are some limits to the financial return the university can expect because the company will retain a correspondingly high proportion of revenues. In the bioscience field, creation of new spin-out companies has tended to be a more popular approach than licensing to existing companies, although in recent years licensing has been increasingly pursued. With licensing the financial return tends to be lower, but universities are accepting that this can be balanced by the reduced time in which it can be secured, and by the far lower chances of project failure.

The large commercial potential of bioscience IP is such that it is very attractive to financiers seeking opportunities to grow their capital by investing it in enterprises that will expand. In addition, biotechnology applications are often so new and unusual that they do not fit into the technical or commercial capabilities of existing companies. Lengthy periods of continued research and development are often required (in the case of new drugs this may be many years) before new products and services are ready for market. Newly created, highly specialised spin-out companies are well suited to commercial projects with these characteristics. Also, because the academic institution and possibly the researcher responsible for the technology will usually own a large share of the company, the spin-out approach can produce a large financial return to these stakeholders if the project is commercially successful.

Small and medium-sized companies have therefore become a core part of the commercial biotechnology sector, many of them spin-outs from academic research departments. These companies may take their products to market themselves. However, it is very common, particularly in the new drug and medical devices sectors, for the technology either to be integrated into new products developed by a larger company, or for the larger company to take on the later stages of development. The larger company then manufactures, markets and distributes the eventual product under a licensing agreement.

1.5 – The reality of risk

Commercial work is inherently risky and you only need to consider media reporting of industry to be aware that economies and markets are always at risk of decline, and that companies all too frequently encounter a variety of threats to their survival. Survival is not too dramatic a term to use: when things go wrong with companies, complete financial failure is a very serious risk, with obvious consequences for the people whose jobs are at stake and for the prosperity of their local communities. Commerce has a schizophrenic characteristic of generating wealth and security on the one hand, and economic devastation on the other, largely because the environment and operational conditions that companies face is fluid and constantly shifting. Factors that cause rapid and dramatic changes in the ability of companies to trade profitably include:

- the emergence of powerful new competitors with newly developed and more attractive products and services, or with similar ones that are cheaper because of production in developing countries or due to lean engineering or production

- fluctuations in rates of interest for the borrowing required to fund development and expansion
- fluctuations in rates of exchange for foreign currency required to buy essential raw materials, and for foreign customers to buy the goods and services
- managers whose experience and training leaves them ill-equipped to understand changing market and trading conditions, or to handle complex financial needs

Science-based companies are as subject to these challenges as any others. But they are also faced with additional, often formidable problems encountered in the developmental stages of the company or product; problems associated with the actual process of technology transfer.

Technology transfer, by its nature, is usually a progressive development process. Whether a project involves the creation of a spin-out company or the licensing of a technology for an existing company to take to market, there is usually a need for continued development to translate the technology into the product or service. Technical development may be linked to other commercial processes such as sales and marketing (for example, in pursuit of deals with pharmaceutical companies to manufacture and distribute new drugs once their development and testing are complete) and the process differs in both its objectives and approach from the original academic research upon which it was based. The near-market technical development therefore often needs to be transferred from an academic department to a commercial environment where it can be under the direction of commercial management. So, commercial technology projects often begin not, as one might expect, with immediate sales of the product, but with sometimes lengthy periods of development taking place within the spin-out or licensed company. This is an unusual industrial management situation; one involving a commercial entity that is unable to trade until it has completed a long period of activity. Obviously, this is a risky position to be in; there is always the possibility that the eventual product may fall short of the hoped-for capability, and this operational model presents a challenging financial need.

When trading, companies obviously receive revenue from sales. Research and development can be funded directly from sales, or by commercial loans levered on the expectation of future sales and for which current sales can pay for the interest. When trading cannot take place until technical development is complete, research and all other operational costs must be funded by speculative investment. This will usually come predominantly from venture capital organisations, wealthy 'business angels' and/or government funding schemes implemented to encourage business development, or by investment from a larger company with which a commercial partnership of some kind has been established (see *Chapter 8* for full details of how new businesses are funded). *Figure 1.3* illustrates the financial imperatives that technology commercialisation companies face.

Figure 1.3: Chart illustrating the typical cumulative cash profile of a developing technology-based spin-out company.

Its cash position becomes increasingly negative during the period when operational spending exceeds its income. Operating with a negative cash flow is made possible by investment, with the area under the negative portion of the chart representing the amount of investment funds that must be secured. ▶

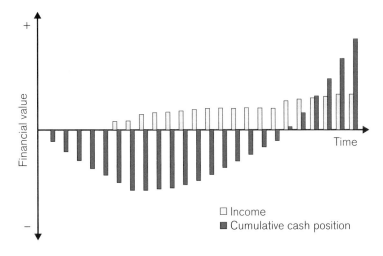

The chart shows how, as the company spends on development costs when its income is absent or limited, the cumulative effect is an increasingly negative cash position. The negative position can be reversed if revenue from sales is eventually sufficient. The area under the negative portion of the cumulative cash position is the value of the investment funding that the project needs. Provided that the development programme can be completed within the expected time and cost, and enough sales can then be generated, the project will become financially secure and there will be a financial return on the investment made. When this is not the case, and key development milestone achievements are not achieved, the resulting cash crisis seriously threatens the project, and the situation can usually be saved only by securing additional investment.

University spin-out companies are particularly prone to cash crisis problems, and there has been an unfortunate history of good companies failing, not because of any problem with the potential of their technology and products, but because they have run out of cash before their technical development is complete, or before they have secured partnerships with larger companies required to take the product to market.

The risks with technology commercialisation, then, are often associated with the relationship between investment funding, route to market and timing. The level of risk closely reflects the commercial sector to which the project is targeted, particularly in the case of commercial bioscience (see *Figure 1.4*).

Some sectors have both very long development times and high costs. This is particularly the case with new drug development, mainly because of

**Figure 1.4: The
development times,
risks of commercial
project failure, and how
these are balanced by
potential capital gain,
for different areas of
commercial bioscience
activity.** ▶

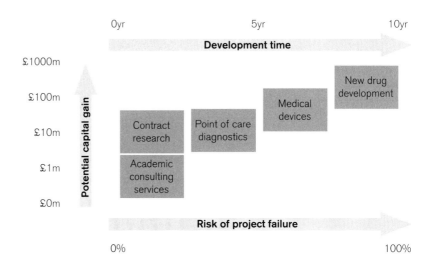

the mandatory need for safety testing and long clinical trials. The huge costs and length of time these take present huge investment needs for small companies. This, and the specialised expertise that is required, means that it is very difficult for a research organisation or spin-out company to take new drug candidates through clinical trials on their own. The usual approach is therefore to seek collaborative partnerships or license the candidate drug to a larger company as its development gets to the later stages. However, there is huge competition from other researchers in universities and elsewhere for deals with the major pharmaceutical firms, and this contributes additionally to risk. The risk is balanced against the large returns (the 'capital gain') that can be secured from investment in drug development when it is successful.

Figure 1.4 shows how other areas of commercial bioscience come with lower risks, often because of their shorter development times and the consequent speed with which sales can be secured from products in these areas. Lower risks, however, often come with lower potential financial return. This balance between levels of risk and likely financial return, and the time over which it can be secured, is a key issue in strategy for technology commercialisation. *Case study 1.3* describes how the success of a young bioscience-based company came partly from a strategy of minimising commercial risk by focusing on bringing the first products to market in a comparatively short time, and with limited development costs.

1.6 – Success factors – the marketing and management imperatives

If we accept a need to plan for risk, it follows that there is also a need to plan actively for success. Success does not necessarily emerge even from

Case study 1.3: Provexis plc
Commercialisation of intellectual property with reduced risk

The bioscience company Provexis was created to commercialise several pieces of IP developed by the Rowett Research Institute in Aberdeen. The Rowett Institute is renowned as a centre for nutrition research, and much of the IP licensed to Provexis has emerged from food technology. This food derivation was the basis of a key opportunity in the company's development strategy because it enabled products to be developed as nutraceuticals.

Nutraceuticals are foods or components of foods that may provide medicinal or health benefits including the prevention and treatment of disease. Importantly, the nutraceuticals field has a sound scientific basis in that the claims of medical effects are or can be supported by laboratory research and clinical trials. It is common for nutraceuticals to take the form of food extracts or active chemicals identified and purified as the bio-active component. Although such extracts resemble pharmaceuticals in their purified form, an advantage of their application is that there is existing confidence in their safety and efficacy because of their history of consumption as food. This means that the safety testing and efficacy trials required prior to their being authorised for sale are generally far less demanding and extensive than those required for a formerly unknown and uncharacterised pharmaceutical molecule. Provexis took advantage of this in its first phase of product development.

Formed in 2000, the company immediately focused on tomato extracts that were effective in inhibiting aggregation of blood platelets, the medical implications of which are substantial reductions in the risks of stroke and heart attack. The commercial potential of this was sufficiently important to generate initial investment funding, enabling the company to undertake an intensive research and development effort in purifying, identifying, characterising and testing the active components. The resulting technology and IP was developed into a product called Fruitflow®, positioned as a natural alternative to aspirin for circulatory disease problems. Fruitflow® was successfully commercialised as a fruit juice, marketed as an over-the-counter 'functional food' that inhibits platelet activation within 3 hours of consumption. Through this approach, Provexis managed to bring a health product to market within the comparatively short period of 5 years, and achieved the enviable position, for a bioscience company, of having a product that generates sales. This first product allowed the company to test the response of consumer markets, guiding subsequent evolution of a more mature product range in partnership with major companies operating in the food and beverage sectors.

The potential of Provexis' emerging strength in the growing market for functional foods, and the talent of its management and staff in developing and positioning Fruitflow®, were recognised in the success of the company's flotation on AIM (London's Alternative Investment Market) in June 2005, raising over £5m of funds to use in its further development.

knowledge transfer projects with the best market potential; there are a number of strategic and operational factors closely associated with success, and they must be proactively considered and built into management.

Market-related success factors are extremely important and include:

- strict market focus – ensuring that knowledge and technology transfer projects are truly market-relevant; creating products and services for which there will be demand and for which people will pay
- market research – a crucially important exercise that must take place at the earliest stages in order not just to verify demand, but to maximise the

match between the eventual product and its end-users, and to quantify the need to enable detailed projection of acceptable pricing, likely volume of sales, and the implications for financial planning, manufacture and other aspects of operational planning

- sales and marketing activities – even the best products and services rarely sell spontaneously. Marketing, of which advertising, public relations and selling are components, must be thoroughly planned and carried out by people with knowledge and experience in the field. It is important to understand that marketing is a major and complex discipline in its own right, and an issue in the knowledge transfer field is the importance of communication between the technical and marketing activities – marketing professionals with technical backgrounds are comparatively rare. (See also *Chapters 4, 6* and *7* and more general texts in the Additional resources section for a discussion of wider market and marketing-related issues and activities.)

Financial management is also a constant issue.

- Funding is crucial to cover the period between completion of development and achieving sales in most projects – project funding is discussed in detail in *Chapter 8*, and is also considered in *Chapters 6* and *7* .
- Operational planning requires detailed projection of all costs and income, including their timing, and legal and taxation aspects mean that financial planning and operation must involve dedicated professionals inside the company and/or the services of competent accounting firms.
- Cash for company operations is amongst the most important issues facing companies on a day to day basis – without cash to pay for things, in both legal and practical senses, they simply cannot operate. Companies frequently fail because they run out of cash. This is common in technology transfer situations where operations are funded by investment and the firm fails to secure turnover from sales in time.

The most crucial component of planning for success, one common to all critical success factors, is management. Experienced professional business managers bring knowledge of:

- markets, sales and marketing
- investment and financial management
- strategic planning
- specialised sector operations and project management skills enabling them to understand, control and reduce the effects of risk

Strategic and operational planning is a highly important success factor precisely because of its effect on the management of risk. It is a common misconception that business plans are prepared as a tool to obtain funding. This view is dangerously wrong. Business plans (see *Chapter* 7) are detailed operational descriptions to enable activities and financial needs to be understood and continually managed. They are based on experience, analysis and projection so that unknowns and consequent risk are minimised. In order to produce business plans of the required quality, to protect IP, and to secure finance to initiate new projects, management expertise is required at the very earliest stages of commercialisation, long before linkage between the academic institution and the commercial world can be established. For this reason, a specialised infrastructure of advice and support has emerged and this is described in detail in the following section.

1.7 — The role of business support

The process of transferring knowledge to generate commercial success requires a wide range of different support skills that complement those offered by the academics and entrepreneurs who have come up with a novel idea. Sound external advice can influence a commercialisation project, making it swim rather than sink. Nowadays there is a whole community of professionals providing assistance to creators of knowledge and technology ideas in order to enable these ideas to become real commercial assets.

Often what is required is a dispassionate opinion from someone with commercial management experience who may have a stake in the venture and who cares about its success. Such a person or organisation may not have been involved in the conception of the original idea or its early development but has been brought in at a later date and can identify the route to how it can satisfy a market need.

Professionals who have acquired experience and skills in different areas of business can provide specific advice on a particular business issue. For example, assistance may be required to establish IPR, to prepare accounts or raise finance, for market expertise or project management guidance, as shown in *Figure 1.2*. Advice may be free or chargeable and can vary in quality. Indeed in recent years both the professional service and public sectors have responded to the needs of technical and business communities by establishing various grants and loan schemes, along with appropriate personnel to distribute these funds, to develop businesses according to merit and need.

Further help and encouragement for successful innovations can be obtained from known mentors and colleagues who themselves may previously have gained relevant commercial experience. Potentially helpful contacts can be obtained through business networks, which may be local,

Professor Colin Garner
Personal profile of a bioscience entrepreneur

Professor Colin Garner's science has always been focused on practical things. He studied pharmacy at King's College, London, and followed this up in 1970 with a doctorate in biochemical toxicology at University College Hospital, London. Like many young European scientists, he then took advantage of the opportunity to travel, taking a post-doctoral fellowship in the laboratory of James and Elizabeth Miller at the University of Wisconsin, Madison, a move that was to be a turning point both in Colin's career and the field of cancer research.

The Millers were pioneers in the understanding of an important problem in cancer research, namely why such a diverse range of chemicals can cause cancer. This effect occurs principally when reactive chemicals bind to DNA, causing somatic mutations that consequently affect the action of genes involved in the regulation of cell differentiation and division. An important characteristic is that many of these chemicals are not particularly active in their native form, but are metabolically transformed when they pass through the liver. The Millers' research on the biochemistry of this process had practical importance because of the potential danger of these chemicals to humans. At the time Colin became a post-doc there was concern that the level of danger from such chemicals was far from fully known because it was so difficult to test for their effects. Prior to the early 1970s the only way to assess chemical carcinogenicity was by animal tests; a process that takes years, is extremely expensive, and ethically contentious. Colin's work in Wisconsin addressed this problem by development of a technique to measure the carcinogenic potential of chemicals through their toxic effects on bacteria. This technique was fast, far cheaper and, crucially, incorporated the possibility of liver activation into the test using liver extracts and a cocktail of biochemicals. Professor Bruce Ames and co-workers at the University of California, Berkeley, subsequently developed the technique further using bacterial

strains that enabled testing specifically for ability to cause genetic mutation rather than being generally toxic to the bacteria. In this format, what became known as the 'Ames test' enabled large numbers of chemicals to be identified as probable carcinogens, contributing to a far more stringent attitude to controlling exposure to them. The *in vitro* liver action component is an absolutely crucial aspect of the Ames test, because many chemicals are only carcinogenic as a result of hepatic activation; without it, the technique would not have achieved its near absolute ability to identify chemicals as carcinogenic or clear them as safe.

Following this success, Colin spent several years at the University of Leeds, and went from there to the University of York, where he established the Cancer Research Unit in the Department of Biology. Although continuing an academic career, his interests were still very much practically focused, and it was at this point that he launched the first of several successful commercial projects, a company providing a series of rapid genotoxicity tests for chemical carcinogenicity as a contract service. Industrial chemicals is a huge commercial sector because chemicals are used so extensively in manufacturing everything from pharmaceuticals to food and materials. New molecules are always being developed, and exposure to chemicals is continuous. This presents chemicals companies with a serious moral and legal need for toxicological testing on a large scale.

Microtest Research Limited, set up by Colin and the University of York in 1976, provided what were then new and technically specialised tests in response to this market need. Microtest was an early example of a university spin-out, a biotechnology company established before the term biotechnology was even in general use. The company enabled Colin and his co-workers to put to practical use the Ames test and a series of other related techniques they had been involved in developing. In addition to the important impact of its safety screening services in protecting human health, Microtest's financial success, though initially modest, helped fund the York group's continuing basic research, and created a number of management, administrative and technical jobs. Colin's farsightedness and practical ability were eventually rewarded when, recognising its technical strength and the important position it had established in an emerging market, international contract research organisation, Hazleton (now part of Covance) purchased the company to extend its services in 1989. Microtest was sold for approximately £2 million, representing a return on investment of several hundred percent. At the time of the sale, the well known venture capital fund 3i, was a significant shareholder in the company.

Another opportunity closely followed this first success. Colin's research group had been amongst the first to use monoclonal antibodies as a research tool, and over a period of time amassed a collection of antigenic proteins and highly specific antibodies that bound to them. Some creative thinking led to this resource being deployed in the unrelated but important area of product authenticity protection – this area is important because counterfeiting of all kinds of goods is a significant commercial problem. A new company, Biocode, was established in the mid-1980s to provide an anti-counterfeiting service in which goods could be marked using secret combinations of different proteins. The authenticity of goods marked in this way can be verified using reagents that contain the appropriate antibodies. Because the marker antigens are secret and impossible to predict or duplicate, this product marking system is extremely reliable. Now known under a new name, Authentix (see *Case study 1.1*), the company has grown successfully.

Colin is currently Chief Executive of his most recent commercial venture, Xceleron Limited. Formed as a spin-out company from the University of York in 1997, Xceleron utilises an ultrasensitive analytical technology called accelerator mass spectrometry (AMS) to study the metabolic fate of new pharmaceutical and biotechnology compounds during development. AMS was developed in the mid-1970s for radiocarbon dating in archaeology and remained in the domain of nuclear physicists for nearly 20 years. Xceleron pioneered the application of AMS in biomedical research. Xceleron's AMS instrument, occupying the size of two tennis courts, has been used to develop new concepts in pharmacology such as microdose studies. In such studies, trace levels of drug are administered to human subjects to investigate their metabolism without having first to test the drug extensively in animals.

Clearly, Colin has enjoyed a successful career that has combined academic research with a major talent for business development. His experience shows how academic science can create effective products, services and companies when technical capability is creatively linked to market openings. Colin comments that from a personal standpoint he has benefited from bridging the academic/commercial divide. Developing novel scientific approaches so that they become commercial successes has been both a stimulating and sometimes frustrating experience. Being an entrepreneur requires good scientific knowledge coupled with enthusiasm, dedication, hard work, luck and, most importantly, belief in what you are doing. Selling science to companies is little different to selling science to grant-awarding authorities, except that decisions are taken more quickly in the commercial, compared with the academic world.

regional, national or international, generic or specialist. These networks offer opportunities to obtain ideas, information and support, and to identify potential partners and collaborators. Professional bodies such as the BioIndustry Association (www.bioindustry.org) can provide access to events, conferences and training in order to obtain industry-relevant advice.

Nowadays most universities employ staff who are dedicated to technology transfer activities in order to support the commercialisation of locally developed ideas. Science parks, business incubators and local organisations such as Business Link (in England), Regional Development Agencies, the DTI Small Business Service and Manufacturing Advisory Services, often provide free or subsidised support. Moreover, commercial consulting firms can provide a wealth of independent experience obtained from the private sector.

The business relationship between academics and these individuals or organisations may be short term, or ongoing and enduring, as the technology transfer develops to maturity and a product enters and captures a place in the market. Prudence must be exercised about who to consult and where to go for advice but, ultimately, successful technology or knowledge transfer requires that a partnership should be established between individuals with technological abilities and those with commercial expertise.

1.8 – Stakeholders and benefits

This chapter began by highlighting the importance of new industry to the health of national economies. The dependence of economic prosperity on innovation and technological development has led to a close linkage between academia and economic development programmes operated by national and local governments (for example, the business support services provided by regional development agencies described above). But what is it about knowledge and technology-based business that makes it capable of contributing to economies so well?

Technology produces products and services by bridging capability gaps, and so there can be great demand for technical solutions to particularly serious problems. This high level of demand is the key, because the consequent level of sales achieved when this is the case can lead to companies growing very rapidly. Knowledge-based business is therefore important to economies because of its high-growth characteristic, its rapid creation of quality jobs, high levels of revenue and company expansion, all leading to creation of new wealth to circulate in a regional economy.

Knowledge and technology transfer works as far more than a source of new ideas for companies; it generates a number of benefits that, although diverse, are all linked. These include:

- business development opportunities for industry
- sources of additional revenue that provide academic institutions with greater independence and development ability
- financial rewards for academic personnel which, although they will not always be large, are welcome in a sector where salaries are rarely high

- a route to enable researchers and their departments to use their skills and facilities for the potential benefit of humanity and the world
- economic development impacts arising from increased business competitiveness, wealth creation and employment

Research has long been recognised by governments as the primary source of progress in, for example, medicine and engineering. But more recently, governments have actively integrated knowledge and technology transfer into economic policies and programmes. For example, the development of new, knowledge-based industries has helped distressed regional economies recover. Through interventions that encourage the creation and location of academic spin-out companies (by funding support programmes, company start-up finance schemes and business premises), local and national governments have a tool to generate recovery in areas where traditional industry has declined.

A specific and frequent aim of government strategies in the developed world is to encourage the development of regional clusters of contemporary business sectors. Clustering is an economic phenomenon where dynamic interaction between regional concentrations of companies in a particular business sector, their suppliers, and wider supporting organisations such as universities, government agencies and professional services, generates particularly strong synergy, resulting in particular regional strength for the sector concerned. Attempts to create or expand technology business clusters can be effective in regions where there is particular research strength (for example, consider how academic strength has resulted in the emergence of industrial biotechnology clusters around Cambridge and Oxford). A complete description of cluster theory can be found in Porter[3].

Knowledge transfer between academia and industry, when properly managed and with close linkage to market demand, is therefore a powerful tool in the pursuit of prosperity and improvement of the human condition. Whether we are scientists or managers, employed in academia, commerce or the public sector, we all owe it to ourselves and one another to pursue it.

1.9 – References

1. *Beyond Borders: The Global Biotechnology Report, 2006*. Ernst and Young, London.
2. **Brown TA** (2006) *Genomes 3*. Garland Science, New York.
3. **Porter M** (1998) *The Competitive Advantages of Nations*. Palgrave Macmillan, London.

1.10 – Additional resources

Tidd J, Bessant J and Pavitt K (2001) *Managing Innovation*, 2nd Edition. John Wiley and Sons, Chichester.
A highly readable and interesting book on innovation as a science and tool for improvement of organisations and development of new business. It contains many interesting case studies of the commercial application and exploitation of new technology.

www.dti.gov.uk/science/
The UK Department of Trade and Industry website offers an interesting introduction to the government department concerned with the application of science for national economic development. A conduit to knowledge transfer and its current activities. There is an area of the website (www.dti.gov.uk/sectors/biotech) specifically concerned with pharmaceuticals and biotechnology.

www.ey.com
Ernst and Young Biotechnology Reports – a definitive series of annual reports on the condition of the global commercial biotechnology sector. The quality of the research and writing in these reports reflects their origin in one of the world's largest and most respected consulting firms. Reference to any of these for the most recent years provides very good insight into the sector.

plus.i-bio.gov.uk/ibioatlas/contents.html#updateslist
The Biotechnology Regulatory Atlas offers an easy-access guide which maps out the main technical regulation that affects biotech companies, particularly those starting out without access to a full time regulatory affairs department.

www.ktponline.org.uk
Knowledge Transfer Partnerships is a DTI programme helping businesses to improve their competitiveness and productivity through the better use of knowledge, technology and skills that reside within the UK knowledge base. There are opportunities for graduates to gain industry experience by becoming a KTP associate.

www.oecd.org
The Organisation for Economic Co-operation and Development (OECD) is an international organisation focused on economic development that generates many reports on a wealth of commercial and economic issues, including those concerning technology. These are available for download from the website and are an excellent source of data and comment for use in essays and professional reports.

www.worldbank.org
The World Bank is another economic development organisation that has extensive applied technology and related information available on its website.

Learning outcomes

Key learning points from this chapter are:

- The concept of markets being generated by capability gaps between human need and aspiration, and the bridging of these gaps by successful products and services

- Technology as a particularly effective capability bridge, and its consequent ability to access and develop market opportunity

- The high growth capability of well-managed technology-based companies, and their consequent contribution to economies

- How technology can be deployed as a tool of regional and national economic development

- The well-established interactions that now occur between academic research and industry

- The different experience and skills of people from academic and industrial backgrounds, and how they can be strongly synergistic when people and organisations work effectively together

- The absolutely crucial need for successful bioscience commercialisation to be based on identified market need and consequent demand – market pull rather than technology push

- The importance of seeking and following specialised advice when commercialising research, and the importance of academic technology transfer offices and regional economic development programmes as sources of this advice

- The risks inherent in commercial technology transfer, and the importance of strategies to minimise them

Chapter 2
Creativity and Innovation in the Biosciences

David J. Adams & Paul Grimshaw

2.1 — Introduction

Inspired ideas lie at the core of all successful enterprises and entrepreneurial bioscientists are expected to be both creative and innovative. *Creativity* is characterised by original, imaginative thought leading to the generation of novel and useful ideas. *Innovation* may be defined as the successful exploitation of these ideas in the development of new methods or devices. Most individuals have considerable creative and innovative potential, but are rarely encouraged to develop and demonstrate this capacity to be original and inventive. Currently, remarkable advances are being made in many fields of the biosciences and there may never have been a more exciting time to be, for example, an ecologist, neuroscientist, or molecular biologist. This chapter shows you how to generate and explore your own ideas and engage in creative approaches to problem-solving, using approaches and exercises designed to promote creativity in individuals working alone and/or as part of a team.

Everyone has a unique perspective on life. It is therefore possible that when a group of individuals is given a problem to solve, each member of the group may suggest a different solution to the problem. However, in a group setting, the views of the more dominant and extrovert participants often rapidly prevail. This may mean that the potentially useful and worthwhile ideas of the less forthcoming group members are lost during discussions. This chapter describes the generation of ideas and development of creative approaches to problem-solving by individuals, i.e. by all members of a team *prior* to group discussions. Similarly, the approaches described for group sessions are designed to maximise the potential for interaction between *all* of the participants. Subconsciously, everyone considers problems during periods of leisure and so *Section 2.3* describes the importance of including a significant incubation period(s) during idea generation and problem solving.

There is a remarkable plethora of techniques designed for the promotion of creativity, for example, Smith[1] analysed over 170 idea-generation methods. This chapter, however, focuses on just a small number of creativity techniques, although they differ markedly in approach. Certain strategies for promoting creativity/innovation will work better for some individuals than for others; furthermore, bioscientists may find that particular techniques are better suited to some situations and problems than to others. The distinct and varied strategies outlined should help you develop a confident and enterprising approach to idea generation and problem solving. In the context of entrepreneurship, the techniques described are invaluable for scientists seeking inspiration to help them break through the creativity barrier they perceive as they stare at a blank sheet of paper.

2.2 – Promotion of creative approaches in individuals

Scientists are expected to be curious about the world around them, inspired by the unusual and always ready to challenge assumptions – these are important traits of great value in the creative process. The strategies outlined below will help you to adopt a creative approach when tackling specific problems, or issues, such as the identification of a novel application for a new technology, or an innovative approach to crisis management in an industrial or clinical setting.

It is worth stressing that creativity often appears related to the lifting of inhibitions in individuals[2]. 'Disinhibited' people are more inclined to take risks, to consider what might at first appear irrelevant, and to be imaginative. Albert Einstein said that 'imagination is more important than knowledge'. So, have the confidence to take the plunge, consider the unusual, make connections between the apparently disparate, and be creative!

2.2.1   Curiosity and inspiration from the unusual

Essential advice to anyone embarking on a career in scientific research, is that he or she should be curious about the unusual and should be prepared for the unexpected. A good example of the scientific and, ultimately, commercial benefits that can accrue from fundamental, curiosity-driven research is the work undertaken during the last few decades with thermophilic microorganisms. Some of these bacteria can grow at temperatures in excess of 100°C; when first discovered they were of great interest to biologists but had no immediately obvious application. However, thermophiles have proved to be an invaluable source of heat-stable enzymes, including DNA polymerases, essential for the PCR procedure (see *Section 2.2.2*), and a range of proteases, lipases and other enzymes that form the basis of so-called biological washing powders.

Sometimes scientists display an impressive combination of curiosity and stubborn persistence. In 1964, Dupont chemist Stephanie Kwolek was engaged in a search for a new strong but lightweight fibre. She explored novel ways of making polymer fibres and noticed that a new polymer preparation was runny and cloudy. Most of the polymer solutions prepared by Kwolek and her colleagues were clear and viscous, and at this point many scientists might simply have discarded the thin, cloudy material assuming that an error or microbial contamination had occurred. Instead, Kwolek was intrigued and continued to work with the unusual preparation. Further investigations identified additional strange features and Kwolek's persistence paid off: her diligence led to the development of the super-strong, lightweight fibre Kevlar. Kevlar has an extremely wide range of uses and, in its most famous application, is thought to have saved many thousands of lives as the bullet-stopping component of body armour.

Finally, the Pfizer drug Viagra provides a recent example of the benefits and profits that can arise from the exploitation of an interesting and unpredicted development. UK-92 480 (as Viagra was known during development) and related compounds were initially screened in programmes designed to identify novel drugs to treat high blood pressure and, subsequently, angina. Preclinical studies with UK-92 480 were encouraging, but the results from clinical trials were much less promising. However, almost hidden among a range of minor side-effects noted for the drug were some reports of penile erections. The clinical investigators and their colleagues decided to pursue these unusual observations and the rest, as they say, is history.

Louis Pasteur said that "chance favours only the prepared mind". From time to time during your scientific career you will observe unusual phenomena and obtain unexpected results. Wherever possible be sure to pursue these observations and make the most of them.

2.2.2 Challenging assumptions

Young children often challenge well-established ideas and ask lots of questions; during these early years their enthusiasm and inquisitiveness are frequently matched by impressive displays of creativity. Disappointingly, as we grow older, we tend to lose the curiosity of our childhood years and many would argue that creativity is suppressed by our formal educational systems. It is important, therefore, for us all to ask the questions 'why?' and 'what if?' more often. The following problem provides an interesting illustration of how we cling to assumptions. Consider the nine dots arranged as shown in *Figure 2.1* and then try to link them all up using no more than four straight lines and without lifting pen from paper or retracing the lines. Most people assume that the lines must not extend beyond the boundaries set by the outer dots. However, this condition was not part of the original problem! If one discards this the problem can easily be solved (see *Figure 2.2*). The solution, quite literally, involves 'thinking outside the box' or lateral thinking[3].

Figure 2.1: The nine dots problem.
Try to join all nine dots together using no more than four straight lines and ensuring that your pen doesn't leave the paper and you don't retrace any of the lines. ▲

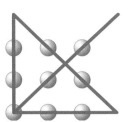

The strategy of challenging assumptions has proved highly successful in creative companies like Dyson and Apple computers. Disappointed by the inefficiency of the conventional vacuum cleaner, James Dyson questioned the assumption that these machines should suck air through bags and filters that rapidly become blocked. As an alternative he designed the highly original and efficient dual cyclone machine that spins dust out of the airstream in transparent bins that allow the user to see when the container is full (for further details, see *Case Study 3.1*). Similarly at Apple, Jonathan Ive asked why computers and screens should be contained within separate dull beige boxes and came up with the innovative and popular i-Mac, with the computer placed inside a coloured, translucent television. Research scientists are expected to question assumptions and this practice has been associated with major developments in biology. For example, Kary Mullis's discovery of the polymerase chain reaction (PCR), during the early 1980s, enabled Mullis and collaborators to challenge the assumption that significant quantities of a specific fragment of DNA could only be produced *in vitro* following cloning and over-expression of the fragment in a suitable host organism. Their work led to the development of a procedure that involves a novel application for enzymes from thermophilic bacteria and the generation of a very large number of copies of a DNA fragment in a reaction vessel (see also *Section 3.4* for details of how Mullis's employers protected this idea using patents). PCR has proved invaluable in a wide range of applications including genome sequencing, medical diagnostics and forensic science.

Challenging assumptions is an important part of being a bioscientist. For example, the diagnosis of most microbial diseases assumes the isolation of samples from patients in hospital wards and the processing of these samples using time-consuming procedures conducted in laboratories that are often located some distance from the patient. It is possible to challenge these assumptions, such that the reverse of this situation would be a simple, rapid diagnostic approach located in the hospital ward. This line of thought might prompt you to consider, for example, an easy to use, near-patient, microbial metabolite detection system that exploits recent advances in biosensor technology. A detection system of this nature is currently the goal of a number of clinical research laboratories worldwide.

2.2.3 Nominal brainstorming and mind mapping

Brainstorming, in which participants' brains are used to 'storm' a problem, is generally assumed to be a group process. Brainstorming in teams is undoubtedly an effective, creative process (see *Section 2.4.1*), but individual, or nominal, brainstorming can also be a useful method for generating ideas and solving problems. The process is simple: you consider a problem and then try to generate as many solutions as possible, usually during a defined period of time (10–15 minutes may be sufficient). Don't be afraid of apparently absurd or impractical suggestions; suspend judgement for the time being. You can return to your ideas and subject them to detailed criticism and assessment at a later stage.

The results of a brainstorming session can be recorded, developed and exploited using the equally powerful mind mapping technique (see www.mind-mapping.co.uk for further details). Mind maps facilitate the connections that can be made during the creative process and the generation of additional new ideas from those already captured in the emerging mind map. To begin, write the problem in the centre of a piece of paper turned on its side and draw a circle around the problem. The example used here is 'antifungal drug discovery'. Brainstorm potential approaches to the problem, e.g. screening for novel compounds, modification of existing compounds, identification and exploitation of novel target sites, exploitation of structural biology, genome sequencing data, etc. Draw lines out from the central circle for each of these potential approaches (*Figure 2.3*). You can now brainstorm each approach in more detail. Thus, you may decide to screen both natural and synthetic sources for novel antifungal agents and this raises further possibilities – these possibilities should be drawn down as additional branches. At this stage, connections between different approaches start to become apparent. For example, you may decide to use information about the structure of a well-defined enzyme target to inform the synthesis of a new class of antifungal agents. All of this can be achieved fairly easily by drawing a mind map manually on a sheet of paper. There are of course a number of software packages that can rapidly create comprehensive mind maps during brainstorming sessions, including the freely available FreeMind; the mind map shown in *Figure 2.3* was created using this software.

2.2.4 Analogies

An analogy may be defined as a comparison made to show a similarity; in a scientific context, a simple example is the analogy between an atom and the solar system. In an analogy, two things that are essentially different, but which nonetheless have some similarities, are compared and this approach can help to promote creativity. Indeed, Koestler[4] considered 'the real

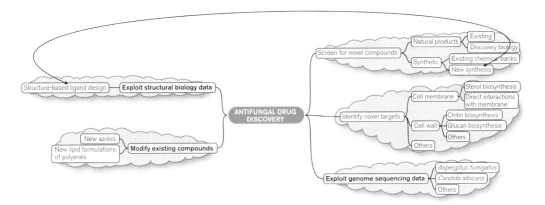

Figure 2.3: A mind map addressing the issue of antifungal drug discovery. ▲

achievement' in many scientific experiments to be 'seeing an analogy where no one saw one before'.

A recent and celebrated illustration of the creative use of analogy during problem solving was the NASA response to the problems encountered with the lens mechanism of the exceedingly expensive Hubble telescope. While taking a shower, James Crocker, one of the principal engineers on this project, reflected on the adjustable and flexible nature of the showerhead. He drew a crude analogy with the structure of the Hubble telescope and proposed a solution (ultimately successful) to the problems that involved placing adjustable/extendable mirrors in front of the malfunctioning lens to realign its focus.

Analogy and bioinspiration

It may be possible to identify an analogy between a problem you are trying to solve and a natural phenomenon. For example, packaging and transport of potato crisps present formidable challenges because loosely packed crisps occupy a great deal of space and break easily during transit. Employees of Procter & Gamble identified an analogy between crisps and dry leaves; the latter rapidly fill sacks and fragment easily when packed together. They also realised that when leaves are wet they are pliable, pack closely together, occupy much less space and remain intact. Based on these observations, Procter & Gamble developed a procedure that involved moulding and stacking wet slices of potato flour which they dried, packaged in tubes and marketed as the highly successful Pringles®.

Strategies and mechanisms adopted by plants or animals can also act as direct and powerful sources of inspiration for ideas. In each of the examples that follow an analogy has been drawn between the role of a particular structure in nature and a potential application for related structure(s) in a commercial setting – this concept is known as biomimetics.

Velcro®. Plants such as burdock and cocklebur produce fruits known as burs as part of their seed propagation strategy. Burs are covered with thin spines with sharply hooked ends that snag almost anything brushing against the plant. In 1948, the Swiss engineer George de Mestral returned from a walk and marvelled at the large number of burs bound firmly to his dog's coat. de Mestral realised that similar hook-and-loop mechanisms manufactured from nylon could form the basis of novel fastenings and Velcro® came to market in 1955.

Lotusan®. The self-cleaning properties of the lotus plant are legendary. With the advent of the scanning electron microscope it became apparent that the waxy surface of the lotus leaf is covered in tiny bumps with cavities between. Rain water runs off this surface very easily and the drops carry with them particles of dirt which exhibit much higher affinity for the water than for the leaf surface. The German botanist Wilhelm Barthlott discovered the basis for this self-cleaning, and realised that the effect need not be con-fined to natural materials like the surfaces of leaves. He demonstrated that synthetic water-repellent surfaces with bumps of the right size would also exhibit self-cleaning properties. These properties have been used to pro-duce a self-cleaning paint, Lotusan®, which has proved to be an impressive commercial success.

Telecommunications. Nature provided inspiration during the nineteenth century, at the very beginning of the telecommunications industry, when Alexander Graham Bell deliberately modelled the telephone on the human ear. Much more recently, attention has focused on photonic crystals which lend vivid colour to structures as diverse as jellyfish tentacles and butterfly wings. Photonic crystal arrays create colours by interfering with incident light and engineers are currently seeking ways to exploit the unique capac-ity of the crystals to handle light by including them in novel optical commu-nication and computing systems.

There are likely to be many more analogies to be drawn between prob-lems facing scientists and potential solutions already out there in the natu-ral world. You may decide to look to nature for inspiration and may be able to identify an analogy between your problem and a natural phenomenon. Alternatively you may discover that a useful analogy stems from an entirely different source such as an inanimate object or an obscure activity. Once you have identified an analogy, apply knowledge or technology from the source to your problem and ask yourself whether this brings a new insight and perspective. You may find that the analogy suggests an entirely novel idea.

2.2.5   Networking and the importance of a 'fresh eye'

The fruitful partnership established between Robert Bunsen and Gustav Kirchhoff provides a famous example of the great benefits that can arise when a fresh eye is brought to bear on a problem. Bunsen, a chemist, used the colours generated by heating a chemical sample in a gas flame as a crude indication of the elements present in the sample. However, he found it difficult to distinguish between flames of similar colours and described this shortcoming to Kirchhoff, a physicist. The latter suggested that such flames could be differentiated by examining their emission spectra through a prism. Working together, Bunsen and Kirchhoff found that each of the substances they examined emitted light with its own unique pattern of spectral lines. This led to the creation of the spectroscope and major developments in the science of cosmology. *Case study 2.1* provides a much more recent example of how interdisciplinary networking can promote creativity.

Professor Eileen Ingham

Eileen Ingham has long-standing links with the University of Leeds; she graduated with a combined degree in Biochemistry/Microbiology and now has a well-deserved international reputation as Professor of Medical Immunology in the Faculty of Biological Sciences. Eileen's research career, to date, has embraced a wide range of approaches and techniques. Her early work on inflammatory responses in acne attracted the attention of Professor John Fisher, a mechanical engineer. John wanted to work with a biologist who could help him understand the immunological basis for the wear, tear and inflammation associated with artificial hip and knee joint replacements. Eileen and John established a collaboration that was to prove particularly fruitful and together they and their collaborators have made major, ground-breaking contributions not only in the field of joint replacements but also, more recently, in the emerging discipline of tissue engineering. This has led to the establishment of the spin-out company Tissue Regenix that focuses on the manufacture of acellular natural scaffolds for tissue regeneration. Eileen is quick to stress the importance of collaboration in all of this. Working with experts in fields as diverse as orthopaedic surgery and knowledge transfer, Eileen, John, their students and postdoctoral colleagues have adopted a truly interdisciplinary approach to problem-solving.

Networking of this sort, which may involve cooperation across disciplines, is as important for creative approaches to problem-solving and idea generation as it is for other aspects of enterprise and entrepreneurship. Whenever possible, share your problem with people from backgrounds that differ from yours. Family members, friends studying other subjects, or contacts from the commercial world, are all likely to see the problem with a fresh eye and may suggest novel solutions. Also bear in mind the great

Case study 2.1: UltraCane

The UltraCane, a powerful new mobility aid for vision impaired and blind people, provides a remarkable illustration both of bioinspiration and the creativity that can emerge following the interaction of 'strangers' with markedly differing backgrounds and expertise. Professor Brian Hoyle, an engineer with expertise in intelligent sensing, Dr Dean Waters, a biologist and expert in bats, and Professor Deborah Withington, a biomedical scientist with expertise in human aural physiology, got together to discuss potential collaborative projects. Their interdisciplinary brainstorming gave rise to a novel, ultrasound-based navigation system inspired in part by the capabilities of the bat.

The UltraCane (*Figure 2.4*) employs the 'echolocation' technique used by bats to detect and avoid objects during flight. Ultrasound transducers located near the handle and tip of the cane emit to the front, left, right and upward, and echoes from objects are processed by a computer, enabling estimates of the position and proximity of these objects. Direction and distance information is passed to the user through a clever innovation involving vibrating 'tactors' that are in contact with the fingers of the hand holding the cane. This enables the brain to build a spatial map of the immediate surroundings.

Trials with the UltraCane have been very encouraging and the device has won a number of

Figure 2.4: The UltraCane in use. See www.ultracane.com for further information. ▲

awards. It is estimated that, worldwide, there are 180 million people with impaired vision. At a cost of only a few hundred pounds, the UltraCane should provide many of these individuals with an affordable opportunity to achieve mobility, independence and social inclusion.

potential of new communication and collaborative technologies such as 'blogs' and 'wikis' [5]. By posting your ideas on websites of this type you will potentially engage a large number of people from well outside your usual circle of contacts. Interactions with these individuals may allow you to obtain valuable feedback and help you develop your thoughts and ideas.

2.2.6 Synonyms

Synonyms are different words with identical or similar meanings. These alternative terms can lend a valuable new perspective to a problem. For example, you might be asked to develop a new detector to be used in the diagnosis of a particular disease. A synonym of 'detector' is 'sensor'. When you consider *sensing* the presence of materials associated with a disease you might think of animals that sense the presence of small amounts of materials and in particular of bloodhounds that detect tiny amounts of volatile compounds using their sense of smell. This can lead to consideration of the construction of an 'artificial nose' for detection of specific volatiles associated with a particular illness. This interesting approach to diagnosis of disease is currently being pursued by a number of research laboratories. Note that in this example the creative process is aided by the use of both synonym and biomimetic strategies. A thesaurus is, of course, the best source of synonyms and you will find the thesaurus attached to your word processing package particularly useful in generating alternative terms.

2.2.7 Personal analogy/empathy

Perhaps the most famous example of this strategy was its use by Albert Einstein who, as a teenager, asked himself "how would the world appear if I were to travel on a beam of light?" He continued to ask key questions, such as "would it be possible to travel faster than light?" and his empathetic approach eventually helped lead him to his special theory of relativity. Higgins[6] describes how personal analogy can be useful during problem-solving in industry. For example, he relates how research scientists successfully developed a new reflective window glass product by envisaging themselves part of the molecular structure of the glass and asking 'what has to happen to make us reflective?' It may be worth trying this method by imagining yourself as part of the object or problem under consideration and looking for novel insights and potential solutions revealed by this approach.

Being creative is hard work and sometimes no matter how hard you try, you may find that you can't come up with any useful ideas. Don't panic! There are many more idea generation techniques out there. First of all try to purge all obvious ideas from your mind by writing them down and mentally dismissing all of these suggestions. Now try to come up with some really original thoughts. If this fails you may wish to try a range of procedures that depend on the creation of a forced relationship between your problem and a random word or piece of information[6]. *Sections 2.2.8* and *2.2.9* give two examples.

2.2.8 Clichés and proverbs

Clichés and proverbs promote strong visual images and, when applied to problems, can trigger useful new thought processes. The phrases in *Table 2.1* are among the most powerful for evoking visual imagery[7].

Table 2.1: Examples of clichés and proverbs used to evoke visual imagery

When the cat's away the mice will play	The early bird catches the worm	Like father, like son	Kill two birds with one stone	Don't count your chickens before they are hatched	If the shoe fits, wear it
Monkey see, monkey do	A man's home is his castle	The bigger they are the harder they fall	Birds of a feather flock together	Two's company, three's a crowd	You can lead a horse to water but you can't make it drink
Don't cry over spilt milk	Two heads are better than one	We're all in the same boat	Never bite off more than you can chew	One bad apple spoils the barrel	Put on your thinking cap
You can't teach an old dog new tricks	You can't tell a book by its cover	When it rains, it pours	Don't rock the boat	Too many cooks spoil the broth	Look before you leap
A penny saved is a penny earned	Put your best foot forward	April showers bring May flowers	Don't beat around the bush	Sink or swim	Don't cross a bridge until you come to it

The potential of this technique is best illustrated using an example. Consider a major aim of this chapter – 'encourage individuals to think creatively'. If this is the 'problem' to solve then some of the clichés and proverbs appear of little value. However, others can usefully suggest ideas and approaches that may be taken, for example:

- *when the cat's away the mice will play* – allow individuals freedom to work alone and in unsupervised groups to enable building of relationships between peers and uninhibited freedom of expression in the absence of supervision by a 'teacher'
- *the early bird catches the worm* – encourage work on creative approaches early in the morning, perhaps immediately after waking, when the brain is often more attuned to creative thought
- *kill two birds with one stone* – ask individuals to try working on two problems simultaneously; ideas from one project area may provide random stimulation for the other
- *don't count your chickens before they are hatched* – advise individuals to allow a period for incubation: tell them to nurture their ideas and allow them to develop to the point where they are ready to be 'hatched'
- *monkey see, monkey do* – encourage a wide range of creative approaches
- *a man's home is his castle* – ask individuals to try various approaches in the privacy of their own homes!

- *the bigger they are the harder they fall* – encourage criticism of the status quo: ensure individuals are not afraid to question assumptions
- *birds of a feather flock together* – try to avoid groups of like-minded individuals during team exercises; encourage group diversity
- *don't cry over spilt milk* – encourage individuals to 'fail forward' [8]; the Nobel prize-winning physicist Richard Feynman said "to develop working ideas efficiently, I try to fail as fast as I can"
- *two heads are better than one* and *we're all in the same boat* – emphasize great benefits associated with <u>team</u> creativity exercises

In addition, bear in mind that 'practice makes perfect' and 'if at first you don't succeed try, try again'!

2.2.9 Organised random search/'Googlestorming'

A traditional approach to 'organised random search' is to choose a page of a dictionary at random and use words from the page to help generate ideas. In an alternative and more up-to-date version of this approach you identify, at random, an item in your flat, at university or elsewhere and do a Google™ or other form of internet search on that item. The results can then be used to help promote ideas. In the following example the problem is 'how can we encourage communication between research laboratories in a university?'. The object chosen at random was a printer and when 'printer' was entered in Google™ the following results were amongst those obtained from the first screen displayed:

- printer driver, modem driver
- printer friendly
- printer **working group**
- InkJet, LaserJet
- **full colour printing**

The two highlighted results were then chosen and the descriptions used to generate new perspectives about the problem as follows.

- 'Working group' suggested the establishment of an informal grouping of researchers, who use similar approaches and techniques, to exchange experiences and, perhaps, resources. A separate technology might be discussed each week and workers could share best practice and troubleshooting advice.
- 'Full colour printing' suggested the creation of a newsletter to include pictures of researchers, their interests, preliminary results and other relevant material.

These results were obtained at the first attempt. Frequently, however, it will prove necessary to search using several random terms before useful results are obtained. Persevere and you will hopefully generate valuable insights into your problem.

In this section, only a small selection of the many methods that may be used to help promote creativity in individuals has been considered. Additional approaches are described in a range of publications and a rapidly growing number of websites (see *Section 2.7 – Additional resources* for further information). It is a good idea to try out as many techniques as possible as certain approaches will work better for some individuals (and situations) than for others. Whichever strategy you adopt, it is likely that the creative process will be enhanced if you allow an incubation period during which you can generate, hone and refine your ideas.

2.3 — Incubating ideas

Research scientists and other problem-solvers frequently have ideas and begin to identify solutions to problems at times and locations remote from those normally associated with the laboratory and workplace. Periods of 'relaxed attention'[9,10], in the shower, during sports or other recreational activities, or even during sleep, appear invaluable in allowing individuals to sub-consciously consider the problem in hand. Here are three well-documented examples of the apparent importance of sleep and dreams during the creative process:

- During the first Gulf War, the Pentagon asked DuPont to increase pro-duction of Kevlar fibre used in the manufacture of body armour (see also *Sections 2.2.1* and *2.5.2*). A machine with a key role in the manufactur-ing process broke-down, costing the company $700 a minute and threatening the lives of soldiers on the battlefield. A large team of engi-neers was assigned the task of fixing the machine, but they were unable to locate the source of the problem. However, one member of the team, Floyd Ragsdale, had a dream in which he saw both the water-conduct-ing tubes of the machine and springs. The dream helped him realise not only that the fault in the machine was due to the tubes collapsing (some-thing that was undetectable from their external appearance), but also that the problem would be solved by the insertion of springs to prevent collapse of the tubes. His solution was an immediate success; his dream is likely to have saved many lives and DuPont $3–4 million.
- The nineteenth century German chemist Friedrich Kekulé appears to have enjoyed many creative moments during periods of reverie. Most famously, while dozing at the fireside, he dreamt about the benzene mol-ecule. Benzene was known to contain six carbon and six hydrogen

atoms and there seemed to be no way in which it could be given a chemically intelligible linear structure. However, in his dream Kekulé envisaged benzene molecules as rows of atoms twisting in snake-like motion and he imagined one of the snakes seizing its own tail; the structure then whirled before his eyes. This insight led to Kekulé's theory for the ring structure of benzene and the possibility of the new science of aromatic chemistry.

- In 1921, the Canadian surgeon Frederick Banting worked with American physiologist Charles Best on the cause of diabetes. It was clear at this time that the disease was due to a pancreatic deficiency. Furthermore, it was assumed that the islets of Langerhans, a cluster of pancreatic cells that degenerate in diabetics, normally secrete a substance that regulates sugar levels. However, attempts to treat diabetic dogs with pancreatic tissue failed. Banting believed that this was because other components in the pancreatic extract deactivated regulatory substances secreted by the islets of Langerhans. The solution to this problem occurred to Banting in a dream. He awoke to write down: "Tie off the pancreatic duct of dogs. Wait six to eight weeks for degeneration. Remove residue and extract" (see www.discoveryofinsulin.com/ Introduction.htm). During the degeneration, the pancreas atrophied leaving the islets intact and Banting and Best were able to successfully treat a diabetic dog with an extract from these cells. Subsequently insulin was purified from pancreatic tissue and injections of this compound have saved a great many human lives during the last 80+ years.

All of this suggests that it really is a good idea to 'sleep on it'. Try thinking about your problem as you drift off to sleep; solutions may occur to you during the night or in the morning. Whenever possible, it is worth leaving a gap of a few days between the occasion when you were first made aware of a task and the next session when you will consider the problem, perhaps as a member of a group or team. Inspiration may arrive at any time, day or night and may occur when you are focusing on another activity; ideally try to write down these thoughts as soon as possible after they occurred. During this incubation period, you may also find it useful to discuss the problem with friends, family members and anyone else who shows an interest (see also *Sections 2.2.5* and *2.5*).

2.4 — Generating ideas in groups

The idea generation strategies described earlier are designed to help individuals achieve their creative potential. However, it should be stressed that well-managed, interactive *group* sessions frequently lead to the generation of many novel ideas and the development of innovative approaches to problem-solving. An excellent example is the successful multidisciplinary team approach adopted by healthworkers in Chicago. A large team of experts, with markedly differing roles at the Northwestern Memorial Hospital, worked together to devise effective strategies that resulted in a marked reduction in disease caused by antibiotic-resistant bacteria (see *Case study 2.2*). A range of creativity techniques for groups exists and this section focuses on four key approaches: 'traditional' brainstorming; brainwriting; the 'Lotus blossom' technique; and Edward de Bono's 'Six Thinking

Case study 2.2: Northwestern Memorial Hospital
The importance of team diversity during infection control

The emergence of antibiotic-resistant strains of pathogenic bacteria is of great concern worldwide. Infections attributable to these organisms frequently spread rapidly in hospital environments, with devastating consequences. The successful approach of a US infection control team, in limiting the incidence and spread of drug-resistant pathogens, provides an excellent illustration of the importance of group diversity during problem-solving.

More than a decade ago, clinicians at the Northwestern Memorial Hospital near Chicago were confronted with an outbreak of disease caused by a vancomycin-resistant enterococcus. Their response to this crisis was to build a large team of professionals, with a wide range of expertise, and to solicit the views of 'outsiders' who could provide a unique perspective on the problem:

- Molecular biologists, using state-of-the-art DNA-typing procedures, developed a technique that enabled the rapid identification of a vancomycin-resistant strain of *Enterococcus faecalis*. Crucially, the availability of this procedure enabled the team to track the spread of the organism throughout the hospital.
- Infection control specialists now detected the pathogen on light switches, stethoscopes and other medical equipment. Armed with this information, the team discussed how the

spread of infection might be contained and, most importantly, asked colleagues in other parts of the hospital for their views and ideas. This led to extremely valuable contributions from individuals who had not, until then, been party to discussions on infection control.

- A pharmacist pointed out that over-prescription of antibiotics leads to selection of antibiotic-resistant bacteria and his concerns were acknowledged. This led to a review of antibiotic prescribing policy and a marked reduction in antibiotic use at Northwestern.
- Maintenance officials were recruited to the team and their contribution was the replacement of corridor drinking fountains with washbasins to promote hand washing.
- Admissions and computer staff became involved: they developed software designed to identify ex-patients, returning as outpatients, who were considered likely to re-introduce infection to the hospital.

In all, the establishment of this unique, multifaceted taskforce led to the implementation of a wide range of innovations which together cut Northwestern's nosocomial (hospital-acquired) infection rate to half the national rate. Indeed, a lasting tribute to this unusual team approach is that many of the measures adopted at Northwestern are now commonplace in hospitals in the US and elsewhere.

Hats' method (see *Section 2.7 – Additional resources* for details of other techniques). In each case the creative process is likely to be enhanced by the inclusion of an incubation period during which team members mull over proposals and ideas at their leisure (see *Section 2.3*).

2.4.1   Group brainstorming

Ideally the group should contain 6–12 members including a secretary, who should record all of the ideas generated by the group, and a leader. The leader should inform group members of the nature of the problem to be addressed prior to the initial group meeting. This will allow individuals the opportunity to generate their own ideas/solutions using the techniques described earlier in this chapter. Alternatively, the leader should describe the problem at the beginning of the first meeting and individual group members should be allowed 5–10 minutes to write down their ideas prior to interaction with other members of the group. Individuals can now share their ideas with the group and the secretary should record the ideas on a board or flipchart in full view of group members. The group leader then initiates the collective brainstorming session which may be sparked off by the initial list of ideas and suggestions from individual group members. The session should last for approximately 30 minutes. The group should be encouraged to generate as many ideas as possible; quantity rather than quality is important at this stage. All ideas should be welcomed and recorded by the secretary. The leader must ensure that there is no criticism of any of the suggestions; in-depth analysis of ideas should take place during a later session. Wild and apparently outlandish proposals are to be encouraged as these may lead to truly innovative developments. As ideas emerge, group members are likely to make further suggestions based on the original proposals. The aim is to establish a synergy as members exchange, combine, refine and extend ideas within the group. The group may wish to record and develop ideas using the mind mapping technique (see *Section 2.2.3*).

2.4.2   Brainwriting

Brainwriting is a non-oral form of brainstorming that is particularly useful with a group of shy or reticent individuals. 'Brainwriting 6-3-5', developed by Bernd Rohrbach, is one of several approaches that may be adopted. In this method, the problem is defined at the top of six sheets of paper (see *Table 2.2*) and one of these is given to each of six group members sitting in a circle. Each participant is expected to write down <u>three</u> ideas within <u>five</u> minutes then pass the paper to the neighbour on the right. The neighbour adds his or her variations on these ideas and, after five minutes, passes the paper to the next person who continues to build on the ideas. The process is repeated until all of the participants have contributed to all of the papers.

Table 2.2: *Pro forma* **for the brainwriting 6-3-5 technique**

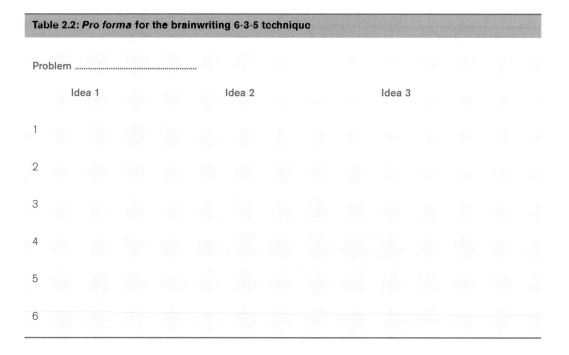

Problem ....

	Idea 1	Idea 2	Idea 3
1			
2			
3			
4			
5			
6			

The group leader then collects the papers and writes the ideas on a board for consideration by the group as a whole.

2.4.3 Lotus blossom

Section 2.2.4 described how the water-repellent, self-cleaning properties of the lotus leaf inspired production of the self-cleaning paint Lotusan®. The lotus proved equally inspirational for Yasuo Matsumura who based the lotus blossom creativity technique on the flowers of this plant. The petals of the lotus blossom cluster around a central core and spread out from that point in ever-widening circles, with one petal leading to the next and so forth. In the lotus blossom technique, a problem, idea, concept or theme is placed in the central square of a three by three grid on a wall chart and the eight cells surrounding the centre are envisaged as the petals of the lotus blossom. Solutions, new ideas, applications, expansions on themes are related to the original problem, idea, etc. and these are placed in the 'petals'. The petals now become the centres of new lotus blossoms (see *Figure 2.5*).

In the following example the technique is used to explore biological applications for liquid crystals. A liquid crystal may flow like a liquid but the molecules of the liquid are arranged in a crystal-like way. Liquid crystals of the sort found in liquid crystal displays can be made to form a complex with biological molecules and are currently being exploited in a wide range of

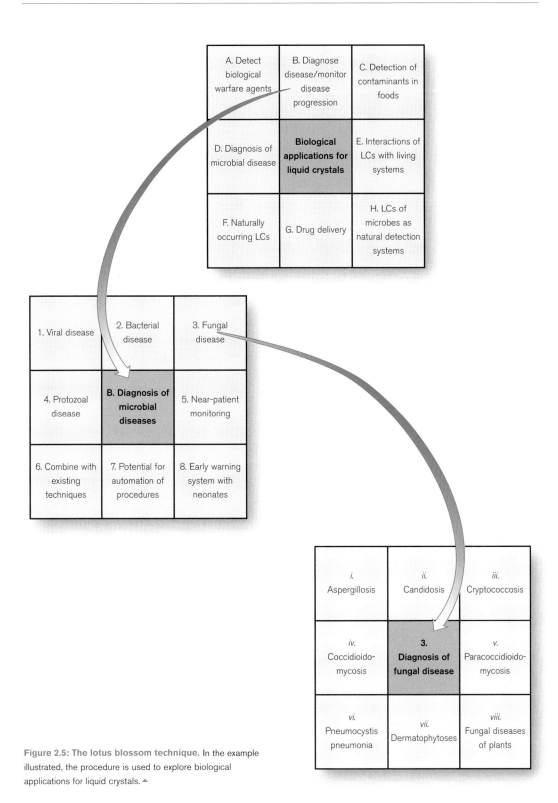

Figure 2.5: The lotus blossom technique. In the example illustrated, the procedure is used to explore biological applications for liquid crystals. ▲

applications ranging from the diagnosis of microbial disease to the detection of biowarfare agents. Most recently, liquid crystal nanoscale matrices have been manufactured as protective cages around delicate therapeutic molecules. Furthermore, it is clear that many naturally occurring nanostructures such as biological membranes are liquid crystals. Liquid crystals are clearly very much a phenomenon and technology of the present and future. The lotus blossom approach provides an excellent framework for consideration of current and potential applications for these materials.

In *Figure 2.5*, the lotus blossom technique has been applied to address the idea 'biological applications for liquid crystals' – this is presented in the centre of the chart. The following eight applications have been identified and are written in the squares surrounding the centre. For each of these applications, additional ideas can be generated, for example:

- detect biological warfare agents – *Bacillus anthracis* (anthrax), smallpox virus, etc.
- diagnosis of microbial diseases – viruses, bacteria, fungi, protozoa, etc.
- detection of contaminants in foods
- diagnose disease and monitor disease progression in affected tissue – assess burns, eye disease, detect tumours
- investigate interactions of extraneous LCs with living systems – DNA, RNA, protein
- research into naturally occurring LCs – DNA, proteins, membranes
- drug delivery – wide range of diseases; LCs in injectable solutions, sprays, topical solutions, etc.
- LCs of microorganisms as natural detection systems? – detect pollutants?

These are then presented in new lotus blossoms. In the example shown in *Figure 2.5* the Lotus Blossom technique is used to further explore the use of LCs in the diagnosis of microbial disease and then, more specifically, in the diagnosis of diseases caused by fungi.

Using techniques like 'brainstorming' and 'lotus blossom', any group is likely to generate a number of ideas and suggestions that it may wish to pursue further. Always try to allow an incubation period of a few days to enable group members to reflect on all of the alternatives. During the next group session team members should be asked to assess collectively the potential and validity of the ideas and to converge on one or more of these. As the group approaches consensus it may prove useful to appoint one of the members as 'devil's advocate'. He or she should challenge the group forcefully and persuasively to ensure that group decisions and plans are robust[11].

2.4.4   Six Thinking Hats®

A group creativity session is most effective if the team includes individuals with diverse backgrounds and expertise. The 'creative abrasion' that can result may be achieved fairly readily in, for example, a large industrial organisation that contains many specialists with markedly different areas of expertise[11]. However, in an academic setting, group diversity may not be achieved so easily as participants usually have fairly similar backgrounds and perhaps a limited amount of knowledge of the material under consideration. There is therefore a need to identify strategies that help academic participants engage with the subject area, adopt alternative stances and interact effectively with one another. A method that allows the group to achieve all of this is the Six Thinking Hats® technique[12] – note that this technique is a registered trade mark of Edward de Bono.

In group sessions, confrontations inevitably arise as individuals with different thinking styles discuss how they wish to approach a particular problem. The Six Thinking Hats® method is a 'parallel thinking' process that minimises conflict by ensuring that everyone is focused in the same direction. Furthermore it ensures that each team member considers a problem from a wide range of perspectives and encourages the participation of otherwise reticent members of the group. With the exception of the group leader/chair person (see below), each member of the team wears the same colour of metaphorical thinking hat at the same time.

- *White hat* – consider the information and data available; focus on the facts, identify and fill gaps in your knowledge and address any questions that need to be asked.
- *Red hat* – covers intuition, feelings and emotions; allows the wearer to put forward a gut feeling with no requirement for justification or explanation; while wearing the red hat you can also think about how other people will react emotionally.
- *Black hat* – the hat of caution and criticism; the wearer is analytical and logical in assessing whether a suggestion fits the facts, the available experience, or policies.
- *Yellow hat* – helps you to think positively and optimistically; look for the good value and benefits in a proposal.
- *Green hat* – the hat of creativity worn when you are generating novel ideas and possible solutions to problems using the wide range of techniques described in this chapter and elsewhere.
- *Blue hat* – thinking about thinking: wearing this hat you review the process and consider what to do next; this is also the hat worn by the person chairing the session who may direct the group to wear one of the other hats as, for example, more ideas are needed (green hat) or the merit of a particular proposal must be assessed (black hat).

The Six Thinking Hats® method can prove particularly valuable in situations where feelings run high and strong opinions are expressed. As an example, consider the response of a group of senior NHS managers asked to devise strategies to enable their health authority to maintain services following a 25% cut to their budget for the forthcoming financial year.

- The chairman starts the session by asking the group to wear *White* hats. They consider all of the facts available regarding the cost of all aspects of the organisation including staff, drugs, facilities, cleaning, acute medicine versus care for the chronically ill, etc. Suggestions start to emerge regarding where cuts can be made but these are not acted upon at this stage.
- The group is now asked to wear *Red* hats and a whole range of emotive issues emerge. For example, group members have strong views for and against cutbacks in cleaning and catering facilities and, more controversially, some argue that smokers or the chronically obese should not benefit from NHS facilities. Others argue that the authority can no longer afford to pay for IVF treatment and a minority propose closure of an entire hospital in an outlying area. These are highly charged issues that all group members consider while wearing red hats. However, when the chairman directs the group to don a hat of another colour they put their feelings to one side and consider the problem from another perspective.
- They are asked to wear *Black* hats to enable a detached analysis of the various suggestions and agree, reluctantly, that they will achieve major savings most effectively by closing a hospital in an outlying area. They agree that they cannot realistically adopt some of the more controversial suggestions such as refusal to treat smokers or the obese.
- However, when the chairman asks the group to wear *Yellow* hats and identify positive, beneficial outcomes that may stem from the various suggestions the discussion on obesity and smoking leads to decisions to promote cheap preventative measures including healthy diets and 'stop smoking' campaigns.
- The chairman now asks the group to wear *Blue* hats so they may review the whole process and decide whether they have enough suggestions.
- They agree that they would like to spend a little more time trying to generate further ideas and therefore wear *Green* 'creativity' hats. During this session someone suggests that the NHS Trust should consider charging patients for small cosmetic operations. This suggestion is considered by the group wearing each of the hats in turn and, ultimately, it is rejected.
- In a final blue hat session the group decide that their response to the funding crisis should be closure of the outlying hospital.

A useful outcome for any of the group exercises would be a 'dragon's den' style of presentation during which the group attempts to persuade a panel of 'experts' of the viability of their proposals and ideas. If a number of groups are involved this can introduce a stimulating competitive edge to the exercises.

2.5 – Entrepreneurs and intrapreneurs

Hopefully, the preceding sections of this chapter will have helped you generate some novel ideas. During the course of your career you will have many ideas and you will want to put some of these into action. If, while working as an individual or with a small group of collaborators, you decide that you wish to exploit the full commercial potential of your ideas then you must be prepared to pursue them with conviction and drive, and you must be prepared to take risks. That is, you must demonstrate a strong entrepreneurial spirit. Alternatively, you may find yourself working within the confines of a large organisation. Under these circumstances you may still find considerable opportunity to exhibit your entrepreneurial skills. It is very much in the interests of employers to encourage their employees to be creative and there are many instances of businesses benefiting greatly from the *intra*preneurial activities of members of staff.

2.5.1   Entrepreneurs

An entrepreneur may be defined as a person who, by risk and initiative, sets up a business enterprise in the hope of profit. During the last two decades, increasing numbers of entrepreneurial bioscientists have created highly successful, profitable companies. This may reflect, at least in part, the numerous schemes that central and local government, and other organisations, have established to provide encouragement and support to would-be entrepreneurs (see *Chapters 1* and *8*). Universities in particular are hotbeds of entrepreneurial activity with the establishment of spin-out / start-up companies on many campuses worldwide.

Entrepreneurial activity in universities is not the sole preserve of members of staff. Many students have shown great initiative in establishing business ventures during undergraduate or postgraduate studies. For example, over 3% of the approximately 35 000 students at the University of Leeds claim to be self-employed in some way (http://ncge.com/files/biblio1020.pdf). The milliondollarhomepage.com website created by 21-year-old Alex Tew provides an impressive illustration of what can be achieved by a determined student entrepreneur. Using a combination of nominal brainstorming and attribute listing (another technique used to promote creativity – see http://www.mycoted.com/Attribute_Listing for further details) this University of Nottingham student came up with the idea of selling pixels, the dots that make up a computer screen, as advertising space at one dollar per dot. The minimum purchase was $100 for a 10 × 10 pixel square to hold the buyer's logo or design. Clicking on the space takes readers to the buyer's website. After four months, Alex had sold space to some 2000 customers and the site had earned around $1 million. As the website mosaic of adverts neared completion an internet icon was created and Alex identified a further enterprising development: selling the image as poster prints!

2.5.2 Intrapreneurs

An intrapreneur is an individual who remains within a large organisation or corporation and uses entrepreneurial skills to develop a new product or line of business. It is in the best interests of large organisations to encourage and nurture the entrepreneurial skills of bright, highly motivated employees. The employees benefit in turn from the assistance and resources that the organisation can provide to bring original ideas to fruition. One of the best known and most spectacularly successful examples of intrapreneurial activity was the development of the Post-it Note by the 3M Corporation. In 1974, 3M employee Art Fry was irritated because his book markers kept falling out and he repeatedly lost his place. He was aware of a non-permanent adhesive that had failed to meet selection criteria during an earlier project. Art placed the adhesive on the back of his markers and found they remained securely in place but could be easily peeled off at a later stage. He saw the great commercial potential of his invention but encountered some initial opposition to its development. Crucially, however, the company continued to provide support and Art was allowed to carry out further research and development. The result was a product that continues to be one of 3M's great success stories. Nowadays 3M has a policy of encouraging all of its nearly 7000 technical employees to work on developing their own business ideas for around 15% of the time they are at work. This enlightened attitude to innovation has helped the company commercialise over 50 000 products; many of these were pioneered by employees who saw a customer need and used company resources to ensure that a product was brought to market.

Intrapreneurial activity need not lead to the development of marketable goods. Intrapreneurs can introduce innovative systems that greatly increase the efficiency of certain practices within organisations. A group of employees in the huge Fibers Department at Dupont, Virginia, was formed to build a database that would enable the company to track the history and quality of every bobbin of Kevlar fiber (see also *Sections 2.2.1* and *2.3*) as it moved through the manufacturing plant. The group was asked to take on additional projects by other parts of the company and, perhaps most notably, by the Medical Products Department on behalf of an important client, the New York Blood Bank. This happened at the onset of the AIDS epidemic and the software proved invaluable in tracking HIV-contaminated blood donations back to source. As demand for the group's services increased, group members were enthusiastically intrapreneurial in establishing a new and highly successful sector within DuPont. The company benefited enormously as the novel tracking technology was used to solve a whole range of problems identified by clients associated with the organisation.

Pinchot and Pellman[13] provide further interesting illustrations of successful intrapreneurs. They also list tactics designed to help an intrapreneur succeed. Here are a few suggestions based on their advice:

- start by modestly explaining your idea to trusted friends; ask them for constructive feedback about its strengths and weaknesses
- within an organisation, you will have colleagues with diverse skills, experience and perspectives; build an internal network of supporters who will help you refine the idea and take it forward
- the organisation is also likely to contain enemies of the idea; try to deal with as many of the potential problems and stumbling blocks as possible before exposing the idea to the opposition
- test the idea, casually, on outsiders and potential customers; this should give an indication of the project's commercial viability
- accept all suggestions graciously and gratefully
- don't give up easily; use negative feedback to help eliminate problems and strengthen your idea

You may find that your entrepreneurial urges are entirely satisfied as an employee if you are fortunate enough to work within a large organisation that nurtures and rewards your innovative contributions. Alternatively, an intrapreneurial corporate culture should provide you with excellent in-house training for a future career as a successful independent entrepreneur.

In either situation, you will find that you must protect your new ideas at the earliest possible stage. In the next chapter a wide range of approaches for safeguarding intellectual property is described.

2.6 – References

1. **Smith GF** (1998) Idea-generation techniques: a formulary of active ingredients. *Journal of Creative Behavior,* **32**: 107–133.
2. **Carson SH, Peterson JB and Higgins DM** (2003) Decreased latent inhibition is associated with increased creative achievement in high-functioning individuals. *Journal of Personality and Social Psychology,* **85**: 499–506.

3. **de Bono E** (1990) *Lateral Thinking.* Penguin Books, London.
4. **Koestler A** (1975) *The Act of Creation.* Pan Books, London.
5. **Butler D** (2005) Joint efforts. *Nature,* **438**: 548–549.
6. **Higgins JM** (2006) *101 Creative Problem Solving Techniques.* New Management Publishing Company, Florida.
7. **Higbee KL and Millard RJ** (1983) Visual imagery and familiarity ratings for 203 sayings. *American Journal of Psychology,* **96**: 211–222.
8. **Maxwell JC** (2000) *Failing Forward: How to Make the Most of Your Mistakes.* Thomas Nelson, Nashville.
9. **Wallas G** (1926) *The Art of Thought.* Harcourt, New York.
10. **Dodds RA, Ward TB and Smith SM** (2003) A review of experimental research on incubation in problem solving and creativity. In: *Creativity Research Handbook*, Vol. 3, Runco MA (ed.). Hampton Press, Cresskill.
11. **Leonard D and Swap W** (1999) *When Sparks Fly. Igniting Creativity in Groups.* Harvard Business School Press, Boston.
12. **de Bono E** (2000) *Six Thinking Hats.* Penguin Books, London.
13. **Pinchot G and Pellman R** (1999) *Intrapreneuring in Action: a Handbook for Business Innovation.* Berrett-Koehler, San Francisco.

2.7 – Additional resources

Bar-Cohen Y (2006) *Biomimetics: Biologically inspired technologies.* Taylor & Francis, London.

Forbes P (2005) *The Gecko's Foot. Bio-inspiration Engineered from Nature.* Fourth Estate, London.
This and the previous book provide up-to-date accounts of a wide range of projects involving bioinspiration.

Boden MA (2004) *The Creative Mind. Myths and Mechanisms*, 2nd edition. Routledge, London.

Simonton DK (2004) *Creativity in Science.* Cambridge University Press, Cambridge.
This and the previous book are for those who wish to find out more about theories relating to creativity in science.

Holyoak KJ and Thagard P (1995) The analogical scientist. In: *Mental Leaps: Analogy in Creative Thought.* MIT Press, Cambridge, MA.
This documents some of the better known contributions that analogy has made during scientific discovery, and in the development and evaluation of scientific hypotheses.

Van Gundy AB (2005) *101 Activities for Teaching Creativity and Problem Solving.* Pfeiffer, San Francisco.
From time to time it is useful to adopt a new perspective when generating ideas and problem-solving; this can be achieved through role play. Van Gundy (and Higgins[6]) describes a range of exercises designed to create situations and open up options that may never occur through sober, analytical reflection.

www.bioscience.heacademy.ac.uk/ and www.hcge.com/
The Centre for Bioscience, Higher Education Academy and the National Council for Graduate Entrepreneurship are national organisations providing excellent resources for enterprising undergraduates, postgraduates, researchers and academics.

www.fbs.leeds.ac.uk/creativity/
This site, for which you will need to register, includes all the procedures described in *Section 2.2,* along with other techniques designed to promote creative approaches in individuals and groups.

www.mycoted.com/
An excellent website, dedicated to the enhancement of creativity and innovation, that provides a repository of useful tools, techniques, mind exercises, puzzles, book reviews, etc.

Learning outcomes

- Creative scientists should be curious about the unusual and should be prepared to challenge assumptions and to think laterally/'outside the box'

- The identification of connections and analogies, between phenomena that have not previously been related, helps promote creativity

- Interdisciplinary collaboration and other forms of networking can provide valuable insights and foster creativity

- Individuals should experiment with 'creativity techniques' until they find the approach(es) that work best for them

- Whenever possible, a period of a few days should be allowed for 'incubation' of ideas

- Carefully structured group sessions should help ensure full and effective participation by all members of a team

- Entrepreneurs should be aware of the benefits of working as *intra*preneurs within large organisations

Chapter **3**

Protecting Ideas

Louise Byass

3.1 — Introduction

In a professional environment colleagues often ask 'What has intellectual property got to do with me?'. The aim of this chapter is to answer this question by exploring the different types of intellectual property (IP) and the ways these can be protected and exploited. Real-life case studies will be used to illustrate some of the points raised and to demonstrate the importance of the issues and the potential impact of getting it right (and wrong).

Although the main emphasis of the chapter will be the importance of IP in the academic or work environment, it is worth remembering that IP also has an impact on everyday life. For example, how many of us have skipped through the copyright notice at the start of a video or DVD without really giving thought as to what the underlying message is, or what it is actually referring to? Whatever your views on the cost of purchasing a DVD or CD, the fact remains that unauthorised copying or distribution, even if from a legitimately purchased copy, is an infringement of the IP rights of the copyright owner, and is against the law. Trade marks are another example of a form of IP that affects us on a regular basis. Many of us like to think that we are immune to consumer branding. It is therefore an interesting exercise to look at a page of trade marks that have had the name of the product removed, and to realise how many are still recognisable. In the supermarket, many of us will go for brands that we recognise in preference to those we are not familiar with, often paying premium prices for the reassurance that buying a known brand gives us. This is usually because we have come to associate certain things, such as quality, with a particular brand, even though less well known brands or a supermarket's own brand may well be just as good. By the use of clever marketing, brand owners can build up huge value and goodwill in their trade marks. As a further illustration of the importance of IP in our lives, consider the regular launch of new ground-breaking pharmaceuticals. It is all too easy to take for granted that a GP can instantly prescribe drugs to make us well, or that headache cures can be bought over the counter in high street stores. However, we should be aware that the pharma companies would be unlikely to spend the millions of pounds necessary to develop these drugs without the protection afforded by filing patents on new drugs and the monopoly that each patent provides.

3.2 — What is intellectual property, and why should I be concerned about it?

Intellectual property can broadly be defined as the product of one's intellect – it is the creative output of the mind of its creator (as Laurence Sterne noted in his novel *Tristram Shandy* in 1761 "…the sweat of a man's brows and the exudations of a man's brains are as much a man's own property as the breeches upon his backside"). This creative output can take many forms, ranging from the artistic, such as a drawing or a musical composition, to the very technical or scientific, such as a piece of computer software or an invention in the field of bioscience.

IP of whatever form is analogous in several ways to other more tangible types of property such as a car or a house. For example, IP can have value. In fact for many organisations, IP is an incredibly valuable asset, sometimes accounting for a large part of the value of that organisation. This is particularly the case for trade marks. Because it has value, IP can be traded in the same way as other forms of property. IP can be sold outright, and this is usually termed assignment. The IP equivalent of renting a house or hiring a car is termed licensing.

As well as having direct monetary value, IP can provide value to an organisation in other, less direct ways. It can enhance the reputation and credibility of an organisation in a particular field, which can in turn lead to other benefits such as generating funding or attracting high calibre staff. If an organisation fails to recognise the importance of IP, or doesn't take steps to ensure it is adequately protected, then this can have a negative impact on the operation of the organisation. An example of this is the fall in share price of a company that is often seen following an unfavourable ruling in the patent courts.

3.3 – Protecting intellectual property

When people think of IP, they usually think first about the forms of IP right (IPR), such as patents, that can be used to protect IP once an idea has been conceived and developed. These different forms of IPR are discussed in *Section 3.4*. However, in order to realise maximum benefit from any IP, it is essential that steps towards protection are taken from the earliest possible point, starting even before an idea has been conceived. In this section, the ways in which IP can be protected at each stage in the development of an idea are outlined.

3.3.1   Before the idea is conceived

Researchers in universities are accustomed to the rather free exchange of results and ideas within academia. Usually it is perfectly reasonable to freely discuss data and ideas with colleagues within the organisation without compromising rights to the information; the university should have enshrined this protection in conditions of employment. Furthermore, when research work is undertaken in collaboration with a company it is done under a formal legal agreement that addresses confidentiality and places restrictions on publication and dissemination of results (see below). However, a real problem for academics stems from their instinct to present their work at conferences and workshops or publish their results in research papers and books. Academics develop their careers through publication of their research in this way but they must understand that dissemination of an idea in the public domain will compromise both their own and their employer's right to gain future commercial benefit from their results and ideas. It is therefore very important that academic researchers consider the need for protection before discussing their data and ideas with anyone other than a close colleague.

There are a number of steps that can be taken, prior to commencement of any project likely to generate new IP, to ensure that your ideas will be protected.

First, look closely at the terms of any funding contract or other agreement, such as a collaboration agreement, that is to cover the work. This may be something that you cannot influence; however, in some cases it will be possible to have the agreement amended before it is signed. In particular look out for clauses that give away all IP, without exception, to another party. Whilst this may seem fair if that party is funding the work, remember that the work would not be possible without the know-how and expertise that you already possess – that is why they have asked you to do the work in the first place. It may not be possible to have the agreement amended so that you own all the IP, but it may be possible to have it state that you will own anything that you develop in a particular area. For example, if methods of analysis are important to your organisation, and new analytical methods developed are likely to prove invaluable for future business, then it may be possible/desirable for you to retain ownership of any new methods developed. At the very least it may be possible for you to negotiate a share of any exploitation income made using the new IP, in recognition of your existing know-how.

If the project involves use of biological or chemical material owned by a third party, it may be necessary to sign a material transfer agreement (MTA, see *Section 3.3.2*) before receiving the material. In such cases, look out for clauses that state that the owner of the material provided will own all the IP you develop using the material – there is usually no sound reason for this. You should also look out for clauses restricting your right to publish the results you generate using the material.

3.3.2 During initial development of the idea

Once you have had an idea or made a discovery, it will usually be necessary to develop this further before it can be formally protected (if applicable) or exploited. For example, if the ultimate aim is to license a particular piece of IP to a third party, you will gain more value from the transaction if the idea has been developed to a stage where it is closer to the market, and also if it has been protected in some way. You should also note that with regard to copyright protection, an idea cannot itself be protected using copyright – it is only once the idea is expressed in some form that it can be protected in this way (see *Section 3.4*).

It is therefore essential that you take steps to protect the idea during the development phase. The following measures are all steps that can be taken to protect IP during development. It is important to note that these measures may also continue to protect components of an idea even after it has been formally protected using other procedures (for example, there is often know-how associated with commercial production of a product that does not necessarily fall under the protection of the patent but that should nonetheless remain confidential). A further consideration is that in cases where formal protection is not possible or desirable, these measures may be the only way of protecting an idea.

Confidentiality

From the moment you have an idea or invent something, the number one thing to remember is to keep it confidential. Although confidentiality of information is implied in certain circumstances, it is best never to disclose the idea or any embodiment of it to third parties unless an agreement is in place that contains adequate confidentiality provisions. This could be an agreement specifically covering confidentiality (see below), or some other agreement with confidentiality clauses.

Confidential disclosure agreements (CDAs), which are also referred to as confidentiality agreements (CAs) and non-disclosure agreements (NDAs), are specifically designed to protect the confidentiality of information that is to be disclosed to others. Often these agreements are between two parties (in which case they can either be two-way, i.e. where both parties will disclose information, or one-way, i.e. where only one party is to disclose), but they can also be between three or more parties.

It is a good idea to have standard *pro forma* agreements available so that a CDA can be drawn up at short notice if necessary. In addition to the standard things that any agreement should contain, such as the names of the parties, general definitions, etc., a *pro forma* CDA should include or indicate the following:

- a definition indicating what is meant by 'confidential information'
- what the information is and what it can be used for – this is often contained in a Schedule at the end of the agreement
- obligations on the parties, who they may disclose to, e.g. employees, and what they are prevented from doing, e.g. disclosing to third parties

- exceptions to confidentiality obligations, e.g. if the information is independently developed by the receiving party, is already in the public domain, or has to be disclosed for legal reasons
- whether information must be returned or destroyed once the agreement has ended
- how long the information should remain confidential from the date of disclosure

It may not always be possible to use your own *pro forma* agreement when putting in place a confidentiality agreement; you may sometimes be asked to sign an agreement originated by a third party. If this is the case, there are certain things to beware of, including:

- clauses that prevent you making copies of the disclosed information
- clauses which state that the information you disclose has to be kept secret by the other party only as long as the agreement is in force – if this is the case and, for example, you disclose something the day before the agreement ends it will only be treated as confidential for one day. It is better if the length of time for which the information has to be kept secret is a certain period from the date you disclose it to the other party, regardless of when the agreement itself ends
- clauses that require you to write down and mark as confidential within a certain period of time (e.g. 30 days from the date of disclosure) anything that is disclosed orally, for example in a meeting. If you then do not write it down and mark it as confidential in the specified time, the other party will not have to treat the information as secret
- clauses requiring you to mark as confidential anything that is disclosed
- clauses that give ownership of any IP, developed using the disclosed information, to the party that disclosed it
- clauses that place obligations on you that do not actually relate to the confidential information, such as provisions preventing you from talking to other companies in that particular field

Even with a CDA in force, care should be taken to protect the confidentiality of the idea. Additional steps that can be taken are as follows:

- make sure that the flow of information is carefully controlled; only disclose to those who have an absolute need to know, and only disclose as much as they need to know – this applies to colleagues as well as to third parties
- it is good practice to keep a record of what is disclosed, to whom and when, in case there are any problems in the future
- confidential documents should be clearly labelled as such
- when not in use, store confidential information in a secure place; do not leave it lying around on desks for any visitors to see

- information that is stored electronically should be password protected
- if information is to be sent electronically by email, make sure that there is a statement at the top of the message informing the reader that the contents are confidential – if the statement is at the end, the reader could already have read and disclosed the information by the time they finish the document and then see your warnings
- make sure that anyone working for you is aware of their duties regarding confidential information

Material transfer agreements

Another type of agreement that may be used to protect certain inventions at any stage of development is a material transfer agreement (MTA). This type of agreement is used when material, often biological (e.g. a microbiological strain or a cell line), is to be made available to a third party. The aim of the agreement is to allow the owner of proprietary material to protect their rights associated with the material by putting constraints on the recipient's use of the material, for example, limiting its use to research purposes only, or preventing the recipient from passing on the material to others. Always put an MTA in place before letting a third party have any biological material, particularly if it is capable of replication.

If you are using an MTA to protect your rights in any material that is proprietary to you, in addition to the standard things that any agreement should contain, the MTA should include or indicate the following:

- details of the material being provided – these are often contained in a Schedule at the end of the agreement; the agreement should also cover progeny and derivatives of the material
- details of what the material can be used for, what it can't be used for, and how it should be used, e.g. according to any applicable regulations
- who owns any IP developed using your material, and if it is to be owned by the recipient, any rights you have to access such IP
- any restrictions you wish to put on the recipient's right to publish results generated using the material
- limitations on any warranties you are providing with the material, and the liability of each party
- whether the recipient must destroy any remaining material once the agreement has ended

3.3.3   During exploitation of the idea

Once an idea or invention has been developed sufficiently, it is likely that you will wish to exploit the idea in order to generate income for yourself or your organisation. In order to exploit the idea, it will be necessary to protect it (or continue to protect it) in some way and there are a number of options open to you at this stage. The way in which any particular idea is protected will usually be decided on a case-by-case basis, depending on factors such as available budget and the nature of the idea/invention.

In some cases where formal protection is either not possible or not desirable, for example, if the idea is not sufficiently novel to patent or where such formal protection is too expensive, you may decide to continue to protect the idea using the mechanisms described above. Thus you may decide, for instance, simply to protect your idea as know-how by maintaining confidentiality. In other cases (usually limited in the bioscience field) the idea may be protected using unregistered forms of IPR such as copyright, or as an unregistered design. Finally, it may be possible to protect the idea using a formally registered form of IPR, for example, by filing a patent application. *Section 3.4* explores these more formal approaches for protection of IP in more detail.

3.4 – Forms of intellectual property rights

The main forms of IPR that can be used to protect intellectual outputs are as follows:

- patents
- trade marks
- copyright
- database right
- designs
- confidence

At first glance the different forms of IPR appear quite distinct. However, there are similarities between some of these and, for certain types of intellectual output such as inventions, there is some overlap between the different forms of IPR that may be used to protect that output. *Figure 3.1* illustrates the stage of a typical invention's life at which each type of IPR can be used to protect that invention.

Some forms of IPR provide a monopoly, whereas others simply prevent unauthorised use of an existing work by a third party. A further distinction between forms of IPR is that some are governed by statute (e.g. patents, trade marks, copyright) and others are governed by common law, e.g. passing off (i.e. misleading the public into thinking that your product is that of another business) and breach of confidence. IPR can also be divided into those rights that require formal registration (generally those that provide a monopoly), and those that exist automatically as soon as the document, drawing, software, etc. is created. It is most often the case that registered

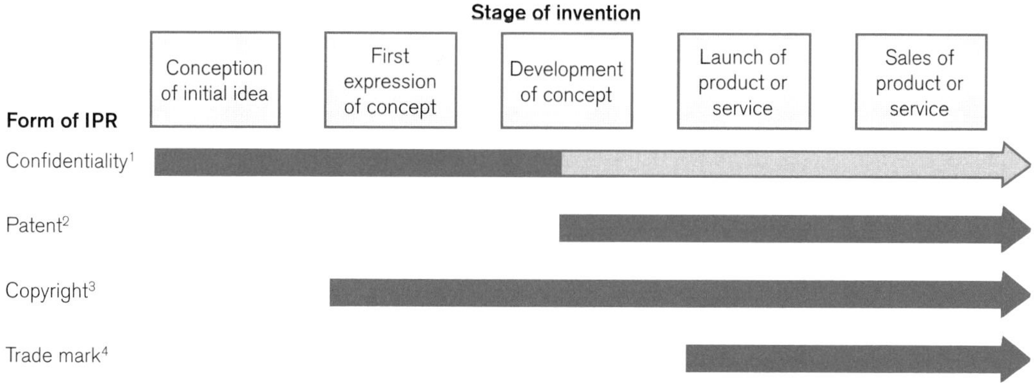

Stage of invention

| Conception of initial idea | First expression of concept | Development of concept | Launch of product or service | Sales of product or service |

Form of IPR

Confidentiality[1]

Patent[2]

Copyright[3]

Trade mark[4]

Figure 3.1: Protection of a typical invention throughout its life, from conception to exploitation. ▲
Arrows represent the duration of relevance of the form of protection; paler areas of the arrow illustrate where that form of protection is still valid but is not as strong.

Notes:

1. Confidentiality should be used to protect an idea or invention from the moment of its conception. If the invention is subsequently patented, this form of protection will only apply to aspects not disclosed in the patent application, such as detail of manufacture or improvements to the invention.
2. Patent application to protect invention itself filed at this point, once the concept has been sufficiently developed, and prior to product launch.
3. Copyright can be used to protect documents associated with an invention, such as drawings for the design of a product. If the product is a piece of software, this may also be the only form of protection available for the product itself.
4. Trade marks may be used to protect any brand name associated with a product or service. An application for a registered trade mark can be made at any time (usually shortly before launch), but the use of unregistered trade marks does not become relevant until significant value and goodwill has been established as a result of use of the mark in the marketplace.

rights are more difficult and costly to obtain and maintain than unregistered rights. For this reason, cost is often one of the major factors that determine which form of IPR is used to protect a piece of IP. Other considerations include the exact nature of the IP itself, the reason why protection is being sought in the first place, and how (if at all) the IP is to be exploited. Exploitation of IP is described in *Chapter 1*.

Sections 3.4.1–3.4.6 outline steps that must be taken to obtain and maintain protection using the most common forms of IPR. Other less common forms of IPR, such as plant breeders' rights, are outside the scope of this chapter, but see *Section 3.5 – Additional resources,* for further information on these.

3.4.1 Patents

Introduction

Patents are perhaps the best known form of IPR, and are used to protect inventions that take the form of products or processes. However, patents cannot be used to protect ideas alone. The grant of a patent by the govern-

Kary Mullis
Inventor of the polymerase chain reaction (PCR)

Kary Banks Mullis was born in North Carolina in the US in 1944. In 1966 he received a Bachelor of Science degree in Chemistry from the Georgia Institute of Technology, followed by a PhD in Biochemistry from the University of California, Berkeley.

In 1979, following a short spell in academia, Mullis began work as a DNA chemist for the Cetus Corporation in California. It was during this time that Mullis invented what is arguably one of the most important scientific developments of recent times. In its simplest form, the polymerase chain reaction (PCR) allows the amplification of specific DNA fragments from complex mixtures using the enzyme DNA polymerase. The resulting highly purified DNA molecules are suitable for analysis and manipulation, and the technique has led to revolutions in medicine, genetics, forensics and molecular biology.

Mullis initially felt that the idea was too simple not to have been developed before. However, initial prior art searches revealed nothing of relevance. Mullis then found it very difficult to persuade colleagues of the potential of the technology. However, his determination eventually paid off and Cetus filed the first of a series of patents protecting the technology. These patents were owned by Cetus as Mullis was an employee of the company. The decision to protect the technology allowed both Cetus and Mullis to benefit from the invention: Cetus earned hundreds of millions of dollars through licensing and assignment of the patents, whilst Mullis also benefited financially from a $10,000 bonus from the company. In addition, Mullis gained significant standing in the scientific community, receiving a Nobel Prize and other notable awards for his work.

Mullis's career continued to flourish after he left Cetus; he now acts as scientific adviser on nucleic acid chemistry to many companies worldwide, and frequently lectures at universities and academic meetings around the world.

ment gives the owner(s) a monopoly right over the invention covered by that patent for up to 20 years from the date of filing. A patent is therefore arguably one of the strongest and most important forms of IPR. The importance of obtaining patent protection for an invention is illustrated in *Case study 3.1.*

As with other forms of IPR, a patent is a negative right. This means that it allows the patent owner to prevent others from making, using or selling the product or process covered by the patent during the monopoly period. However, it does not provide the automatic right for the owner to use the invention themselves – they may be prevented from doing this by the existence of other patents owned by other parties, and this concept (termed freedom to operate) is explored later in this chapter.

A patent is, in effect, a deal between the owner and the state – in return for this strong monopoly right, the owner gives the state and society as a whole a full description of the invention. In the UK, these descriptions are published by the Patent Office. This information, which becomes freely available for use when the patent is no longer in force, provides what is probably a more comprehensive source of technical information than any other.

Case study 3.1: The Dyson Story

The importance of patent protection is demonstrated by the story of James Dyson and his invention of the bagless vacuum cleaner. By understanding the patent system, utilising it to protect his invention, and being prepared to take action to defend his property, James Dyson was able to successfully challenge his rival, the large multinational firm Hoover, when they copied his technology. His story is as follows.

It was in 1979, a number of years after setting up on his own, that Dyson began developing the Dual Cyclone™ bagless vacuum cleaner. Despite the high costs of obtaining patent protection, Dyson recognised the importance of this, and even though at the time he had no income and money was tight, he filed an application for a patent to protect the invention.

Dyson spent two years trying to get large companies to take a licence to the technology, but without success – these companies probably recognised that the introduction of a bagless system would mean the end of the replacement bag market, which was itself worth many millions of pounds per annum. Dyson therefore decided to go it alone.

In 1986, sales of the new cleaner began in Japan. Sales began in the UK in 1993, and by 1995 the Dyson cleaner had become the UK's best selling vacuum cleaner. By the year 2000, sales of the Dyson had caused significant damage to the market share of the rival manufacturer Hoover. Hoover therefore decided to try to win back some of this market share by producing its own range of bagless cleaners, spending several millions of pounds developing its Vortex system. In order to protect its hard won monopoly in the market, Dyson issued proceedings against Hoover for infringement of its patent. Dyson's case was successful, with the judge ruling that Hoover's design was an infringement of Dyson's patent, despite Hoover's claim that the invention was already known in the industry. Ten days later, Dyson won an injunction banning the sale of Hoover's system. In January 2001, a further groundbreaking injunction was granted against Hoover stopping them from re-launching their system until a year after the expiry of the Dyson patent, thus in effect extending the life of the patent. Hoover responded by taking the case to the Court of Appeal, claiming that the patent was invalid because the invention was obvious and lacked novelty, and that the patent itself did not contain enough information to enable the invention to be reproduced (this latter point highlights how important it is that a patent is worded carefully). However, the Court of Appeal decided that Dyson had taken known technology, but had applied it in a new way, confirming that the patent was indeed valid. A petition by Hoover to have the right for a further appeal was unsuccessful.

The final straw for Hoover came when, in 2002, Dyson was awarded £4 million in damages – Dyson had originally offered to settle the claim at just over £1 million in order to avoid long and costly High Court proceedings. Hoover, however, rejected this offer, presumably because they were confident of winning the case. This turned out to be a costly mistake.

By winning this battle of David vs. Goliath, James Dyson has become something of a cause célèbre amongst inventors. But without the protection provided by his patent, he would have been powerless against the infringement by his much larger rival. After being awarded damages, he was quoted as saying "I hope that this encourages inventors who have their ideas stolen by multinational companies to fight for their patent rights". However, the costs of defending a patent can be high and inventors and small businesses often do not have the funds to take on such challenges. Dyson was fortunate that he had the funds to do so.

The main messages to emerge from this story are: patent protection is important if you want to protect an invention such as Dyson's; you must be prepared to fight to defend a patent, even though this can be a costly and lengthy process (you must also bear in mind that, in the end, a patent is only worth what you are prepared to spend to defend it); don't think you can get away with infringing the patents of others, even if they belong to small, independent inventors – this proved to be a very costly exercise for Hoover!

Patents are territorial in nature, in other words patents will only protect an invention in the individual countries where a patent has been applied for and granted. While it is possible to patent an invention in most countries of the world, there is no such thing as an international patent and the owner must therefore apply for a patent in each individual country (although there are some systems in place that simplify this process, and these are outlined in the section *Filing outside of the UK* below).

Criteria for patentability

Because a patent provides such strong protection, there are a number of rigorous criteria that must be met before the Patent Office will grant a patent. These are as follows.

Novelty. To be patentable, an invention must be novel and this means two things. First, that the invention must be different from anything that has been described before, and secondly, that the invention has not been disclosed to anyone else before the date the application is filed. In patent terms, anything that has been described before is known as 'prior art'. This can take the form of scientific papers, existing patents or patent applications, or any other form of publication. When considering whether an invention is novel, it is useful to perform a prior art search. Preliminary searches can be performed using the resources described in the *Patent searching* section below and in *Section 3.5 – Additional resources.* Alternatively, it is possible to instruct a patent lawyer, or other suitably qualified professional, to perform a search on your behalf (although there will usually be a cost associated with this latter option). The importance of prior art is demonstrated by *Case study 3.2.*

The second requirement for novelty is that the invention must not have been previously disclosed (known as 'prior disclosure') anywhere in the world, in any form. Note that disclosure in any form means any written or oral disclosure, whether formal or informal, without exception. This therefore includes a chat over coffee with peers, presentations and abstracts at meetings, as well as the more obvious examples such as a publication in a peer-reviewed journal.

It is important to note that the situation is slightly different in certain other countries, most notably the USA and Canada. In these countries, a 12 month grace period is provided, meaning that a patent application can still be filed following disclosure provided that the application is made within 12 months of the first disclosure.

This issue of prior disclosure is probably the most important matter to remember when considering IP issues. Outside of the USA and Canada, it is usually simply not possible to file a patent application once the invention has

Case study 3.2: Windsurfing International

In 1985, a well-publicised case came to court that served to highlight the importance of novelty of an invention. The case also demonstrated that prior art really can take almost any form.

A company called Windsurfing International was already active in the expanding windsurfing market, and had been granted a patent for the first commercial windsurfer with a particular configuration of sail and board. The company wanted to dominate the market, and did not intend to grant a share of the market to competitors. A rival company, Tabur Marine, saw the potential of the invention, and wanted to get into the market. They therefore started to manufac-

ture and sell sailboards based on Windsurfing International's design.

Windsurfing International consequently sued Tabur for patent infringement. Tabur responded with a challenge that the patent was invalid, and produced prior art that proved this. The prior art was in the form of film footage from 1958, at least 10 years before Windsurfing's patent was granted, that showed a 12 year old boy playing on his homemade version of the sailboard. The court upheld that the patent was not novel and that it had been anticipated, and the patent was revoked. Tabur were then free to make sailboards in competition with Windsurfing International.

been disclosed. It is therefore vital that any publications relating to new work are carefully considered if there is any possibility of a patent application being filed. It is important to note that patenting and publication are not mutually exclusive – once a patent application has been filed, it is in theory possible to publish the invention in a scientific journal, etc. However, take care when publishing, even after a patent application has been filed. For a start, you may want to keep your invention secret from competitors for as long as possible, i.e. until the application is published by the Patent Office (see below). In addition, if you publish improvements to the invention, or describe other uses that were not covered by the original patent application, it may prove impossible to file new applications to cover these improvements in the future.

It is recognised that scientific collaboration often necessitates holding discussions with third parties. If this is necessary, it is possible to hold discussions without jeopardising any patent application, provided that any information disclosed is protected by putting in place a suitable CDA before any discussions take place. This will mean that the invention is protected as confidential information until the application is actually filed. The use of CDAs and the protection of IP as confidential information or know-how was covered in *Section 3.3.2*.

Inventive step. The second criterion for patentability is that the invention is sufficiently inventive. This means that it must not be obvious to a hypothetical person who is 'skilled in the art' – in other words, someone who is technically competent in that field but who lacks the imagination to come up with the idea themselves. In reality, this inventive step does not have to be a huge leap as many patents are granted for incremental improvements to existing technology. Furthermore, an invention cannot simply be a discovery of something that already exists in nature, such as a gene sequence, unless the inventive step relates to its use. Inventive step is very difficult to assess, and therefore decisions on this are best left to the professionals.

Industrial applicability. The rules also require that to be patentable an invention must have some practical use – it is not sufficient that it is merely interesting. Industrial applicability can apply in any field. However, there are exclusions for certain types of invention that are deemed by law never to be capable of industrial application. These include methods of treatment or diagnosis of human or animal bodies (this exclusion does not extend to drugs or other substances that could be used in this way), or new animal and plant varieties.

Not excluded by statute. The final criterion is that in addition to meeting the three criteria listed above, an invention must not be excluded by statute. As well as the exclusions mentioned already, certain other things are excluded from being patentable. Examples include: literary, dramatic or musical works; scientific theories, discoveries or mathematical methods; methods of doing business; computer programs; and inventions encouraging antisocial, immoral or offensive behaviour (such as landmines).

Ownership and inventorship

Where the inventor is a private individual, he or she will usually also be the owner of the patent. However, many inventions are made by employees in the normal course of their work, and in these cases the owner will usually be the employer (although the inventor is still named as such on the application).

It is important that all inventors are listed on the application, as failure to do so can cause problems later down the line. Note that inventorship on a patent application is not the same as authorship on a scientific paper; only those who have had an actual inventive input into development of the invention should be listed as inventors. Others who might normally be included as authors of a paper, such as the head of the group or laboratory, should not be included.

UK application process

There are certain procedures that must be followed when applying for a patent. The procedure and approximate time scales for obtaining a patent in the UK is outlined below, and illustrated in *Figure 3.2*. In reality, most applicants will wish to apply for patent protection in a number of countries, and there are systems in place to make this process easier. These are outlined in the following section.

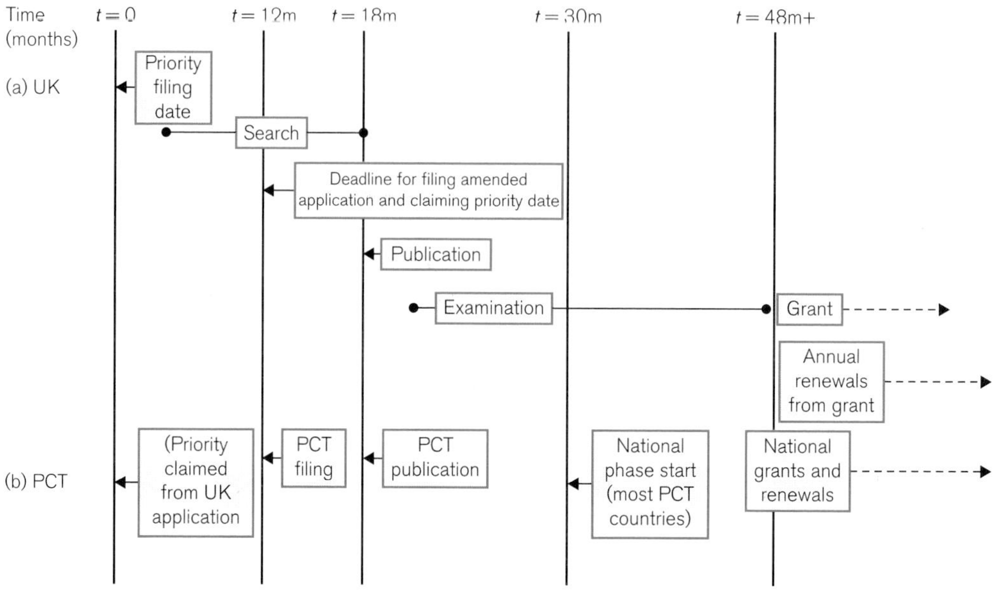

Figure 3.2: Timeline for patent applications. ▲
Outline of (a) UK application process and (b) International application via the Patent Cooperation Treaty (PCT) process (claiming priority from same UK application).

Initial filing. To obtain a patent, the owner must first file an application with the relevant authorities. In the UK, this is the Intellectual Property Office. The date on which this initial filing of the application is made is an important one. It is termed the 'priority date' and it allows the owner to file applications in other countries after this date while still claiming this date as the filing date in these other countries, provided that the applications are filed in the other countries within a certain time (discussed later). In addition, if necessary the applicant is able to file new amended applications covering the same invention during the 12 months from the priority date; any such amended application filed during this period will be able to claim the priority date as the filing date. The filing date is also important because in the event of a dispute, once the patent is granted, the owner can take action against any other parties who are making or using the invention without permission, and this action can be backdated to this priority date. (In IP terms, the unauthorised use of a patented invention by a third party is called infringement.) In the UK, the first to file a patent for any particular invention is granted the patent (provided, of course, they successfully meet all the necessary criteria). In the USA the rules on who will be granted a patent are currently different (see *Filing outside of the UK* section).

Search. Once satisfied that the application is complete (i.e. it has been made on the right form, contains the necessary information and signatures, etc.), the Intellectual Property Office will perform a search of prior art relating to the application (provided that, within 12 months of filing, a formal request for a search is made and the relevant fee is paid) and will report its findings to the applicant. The search report will give an initial indication of how difficult and costly it is likely to be to eventually obtain grant of the patent. If the Intellectual Property Office provides a long list of references that it considers to be directly relevant to the subject matter of the current application, then this could be an indication that it may be a good idea to withdraw the application. If it is withdrawn before publication (see below) the content will remain secret – it will therefore be possible to file another application to protect the invention at a later date because the invention will not have entered the public domain and will still be novel, provided it has not been published or disclosed elsewhere. Alternatively, it may be appropriate to reconsider the content and file an amended application (provided that this is done within 12 months from the priority date).

Publication. Approximately 18 months after the filing date, the application will be published by the Intellectual Property Office. After this time it is no longer possible to withdraw the application and keep it out of the public domain. If the invention has not been published elsewhere between filing and publication by the Intellectual Property Office, it is at this time that your competitors first learn of what you are doing (the titles of all new applications are published by the Intellectual Property Office shortly after the filing date, but patent titles do not usually give very much away). When searching through the patent literature, an 'A' suffix to an application number indicates that this is the version that was published at the 18 month time point.

Examination. If the applicant still wishes to proceed with the application, then within 6 months of publication, they must make a request to the Intellectual Property Office for substantive (i.e. detailed) examination of the application. This is the stage at which the patent examiner will take a closer look to see whether the application fulfils the necessary criteria (novelty, inventive step, industrial applicability). The examiner then informs the applicant of their opinion, and the applicant has limited scope to answer objections raised by the examiner.

Grant. If, at the end of the examination process, the examiner is satisfied that all the criteria have been met, then the application will proceed to grant. This usually occurs between four and five years after filing of the application, and means that the patent is now fully in force and is no longer just a patent application. Upon grant, patents are re-published, and given a 'B' suffix to the patent number.

Post-grant. Once a patent has been granted, the owner can take advantage of the protection they now have for their invention, for example, by licensing the rights to third parties. However, in order to continue to benefit from this protection they will have to pay annual renewal fees to maintain the patent and keep it in force. Failure to do so will generally result in the patent lapsing.

Following grant of a patent, it is possible for the owner to sue infringers. It is also possible for any third parties to start proceedings to have the patent revoked, for example, on the grounds that the invention is not novel or does not have an inventive step. *Case study 3.2* provides an example of a granted patent being revoked following an action from a competitor.

Filing outside of the UK

If the owner of an invention wishes to file for patent protection outside of the UK, then there are systems in place to make it easier to file in each country in which protection is desired. The usual route taken by UK applicants is to make the initial filing in the UK (indeed, UK residents must either file in the UK 6 weeks before filing abroad, or ask for special permission to make an application outside of the UK first – this is for reasons of national security). Before the end of the first 12 months following the priority date, the applicant may take advantage of one of the various treaties that exist worldwide. For example, he or she could file an international application via the Patent Cooperation Treaty (PCT) or file a European application, taking advantage of the European Patent Convention (EPC).

PCT. A PCT application allows for filing in many countries/regions claiming priority from the first application, and the application must be filed within 12 months of the priority date. The list of countries and regions that are members of the PCT is expanding all the time – the current list can be found on the World Intellectual Property Organisation (WIPO) website (www.wipo.int). Using the PCT system, an international application is never actually granted, and individual applications must still ultimately be filed in each country – the system serves only to make the process simpler and to delay the time by which applications must be filed in each individual country (this has the added benefit of delaying the associated costs, giving more time in which to make decisions on how to exploit an invention). *Figure 3.2* outlines the PCT filing process when claiming priority from a UK application.

EPC. Using the EPC system, timescales and costs are again delayed. At the end of the EPC process, a European patent will actually be granted. This must then be validated in each individual European country in which the applicant wants protection; the European patent is then allowed to lapse. It should be noted that the list of members of the EPC is not the same as the list of members of the EU and therefore, as always with patent filing, it is essential that professional advice is sought.

If the applicant chooses not to take advantage of the PCT or the EPC, then an application must be filed separately in each individual country within 12 months from the priority date.

When filing applications in other territories, it must also be remembered that significant differences exist between the UK system and those of certain other countries. For example, we have already seen that in the UK the first to file an application will eventually be granted a patent for any particular invention. However, in the USA, the party who is first to make the invention will be granted the patent (although this is likely to change in the near future, bringing the USA more in line with the rest of the world) – if patent filing in the USA is a possibility, it is therefore essential that documentation (e.g. in the form of lab notebooks) is adequate to prove exactly when the invention was made.

Costs

The costs associated with filing and obtaining of a patent can be significant. Although there is currently no fee for initial filing of an application at the UK Intellectual Property Office (the current application fee of £30, and the £100 search fee, do not have to be paid until 12 months after filing), applicants will usually have the application drafted for them by a qualified patent lawyer. This is because it is important that the application is written in a way that maximises the chances that the examiner will consider it meets all the necessary criteria. When legal costs are taken into consideration, it usually costs a few thousand pounds to file the initial application. There are then Intellectual Property Office fees associated with subsequent steps such as the search, examination, patent grant, annual renewal, etc., together with the associated legal fees at each stage. Over the lifetime of a patent the costs can therefore amount to many more thousands of pounds, and this is for a UK-only patent. Once you start to file in other countries, the costs really start to mount up, mainly because of the costs involved in translating the patent into other languages. Each stage, e.g. examination, renewals, etc. must be paid for in each country in which an application is ultimately filed. The total cost, over the patent lifetime, of protecting an invention in the major territories of the world can therefore be in excess of £150 000. Because of these high costs, applicants must be reasonably certain that they will be able to recoup the costs through exploitation of the invention and should abandon the application process at any time if it looks as though this will not be possible.

The costs of patenting, maintaining a patent and, if need be, defending a patent against infringement can be very high. Also, taking out a patent 'publicises' your invention and unscrupulous operators will think nothing of infringing a patent knowing that a small business is unlikely to have the resources to defend against infringement. This raises very real concerns for inventors and new ventures. Why pay the high costs of patenting if you cannot afford to protect your patent? This is a difficult dilemma faced by many inventors and new ventures. Alternative strategies and decisions are best taken with expert advice. Most universities now employ, or provide access to, qualified patent agents.

Freedom to operate

Earlier in the chapter the concept of freedom to operate was mentioned. Having freedom to operate, when using or making a technology, means that what you are doing or want to do does not fall within the scope of a patent or patents owned by someone else. If it is covered by such a patent, then you do not have freedom to operate and must ask for permission from the owner of the patent to do what it is you want to do. Without this permission you will be infringing the patent rights of the owner(s), and they will be entitled to take action against you. Permission to use someone else's patented technology is usually given in the form of a licence. This will state exactly what you are being given permission to do, where you are permitted to do it, and how much it will cost you – licences usually have fees associated with them, for example a payment when the agreement is signed and/or royalties in the form of a percentage of revenue for as long as the agreement is valid. See *Chapters 1* and *6* for more detail on licence agreements.

Even if you are the owner of a patent that covers a particular invention (e.g. part of the process necessary to manufacture a new drug), it is still possible that by using the patented technology yourself, your actions will fall within the scope of someone else's patent. If this is the case, then even though you own the patent on part of the process, you still need permission from the owners of the other part(s) before you can use your own invention. Consider the simple example below.

- Company X invents the concept of a balloon filled with hot air that will rise and travel above the ground:

- Company Y thinks this is a good idea, but realises that the invention would be improved if people or cargo could be carried beneath the balloon. They therefore invent a basket to hang from the bottom of the balloon:

- However, Company Y's basket is pretty useless on its own, and they cannot make the balloon as well as the basket without a licence from Company X. Neither can Company X add Company Y's basket to its balloon without a licence from Company Y. In these circumstances, one company may grant a licence to the other, or they may each decide to grant a licence to the other – the latter procedure is known as cross-licensing. In this way, both parties can make a balloon with a basket:

Patent searching

You may decide to perform a search to identify existing patents in a particular area for one, or more, of several reasons. For example, you may wish to get an idea of whether your invention is novel, or you may want to see whether you have freedom to operate to use your invention. While patent lawyers will perform such searches on your behalf, this will cost money. There are ways of doing searches yourself for free, even if this is just as an initial 'look-see' when you first come up with an idea. For example, there are a number of websites where patent searches can be performed free of charge and some of these are described below. In addition, *Section 4.5* gives details of sites accessible to the public but where a fee must be paid for the search.

The first website to use when looking for patent information is the Espacenet site (http://ep.espacenet.com). This is operated by the European Patent Office and contains details of over 30 million patents worldwide (i.e. not just European patents). Using this site it is possible to do a quick search using words in the title, abstract, people names, or organisation. There is also an advanced search option where you can search by applicant, inventor, application number, publication date and classification number. The latter can be particularly useful in identifying patents in a specific area of technology; patent titles are usually vague and uninformative and searching by keyword will therefore not necessarily identify all patents of interest. See *Figures 3.3* and *3.4* for screen shots of the search pages.

Figure 3.3: Quick search page of the Espacenet site. ▶

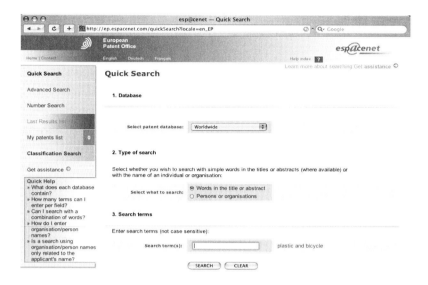

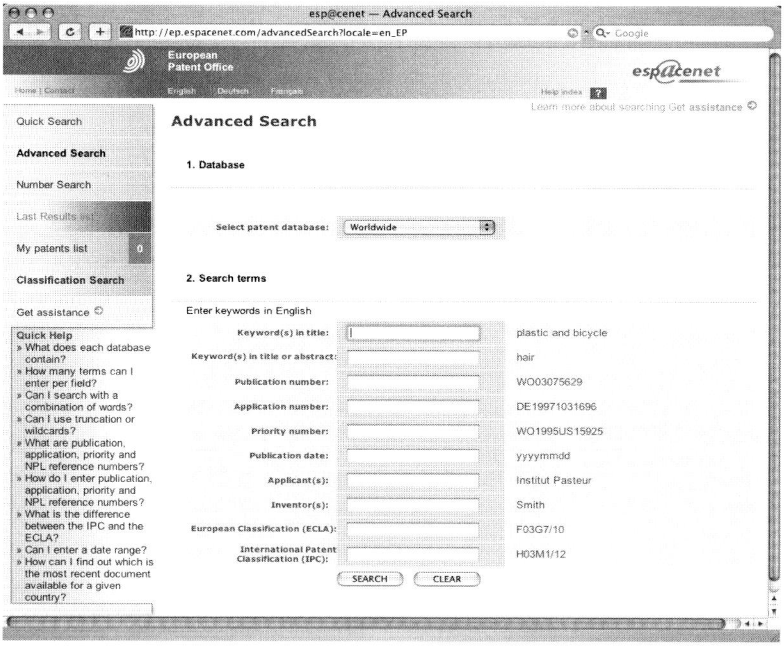

Figure 3.4: Advanced search page of the Espacenet site. ▶

Once the search is complete, a list of relevant patents is displayed. Further details can be accessed by clicking on the link for any of these patents. This includes the full text of the patent, details of other patents in the family (i.e. corresponding patents in other countries), details of the status of the patent (e.g. whether it has been granted, whether the renewal fees have been paid, etc.), and a printable version (as a pdf file) of the original patent application or granted patent (although this can only be downloaded and printed one page at a time). Details obtained using this first search should then be used to perform further iterative searches. For example, if you find a patent of relevance, you may want to look for other patents filed by that company, or use the classification number of the patent to find others in the area of interest.

Other useful and free websites include the US Patent and Trademark Office (www.uspto.gov) and the Japanese Patent Office (www.jpo.go.jp) sites.

Understanding a patent

Once you have found a patent that appears to be relevant, and have obtained the text of that patent, then you will quickly discover that patents are often long, complex and confusing documents that are written in a form of English that is difficult for a non-lawyer to understand. However, understanding the structure of a patent will help you extract the most important information without having to read and understand the whole thing.

The most important parts of a patent are the front page, which contains details of the priority date, applicants and inventors, title and abstract. The suffix of the patent number may provide further useful information – in the UK and Europe, A and a single digit number (e.g. A1) after the patent number means that this is still an application and that the patent has not yet been granted, whereas B and a single digit number (e.g. B1) means that this is a granted patent. If you are concerned that you may be infringing a patent, but cannot find reference to a number with a B suffix, then this may mean that the patent was never granted, and that you are therefore free to use this invention. In such cases it is worth checking with a patent lawyer to find out for sure whether the patent was granted or not. See *Figure 3.5* for an example of the front page of a UK patent application.

Another important part of a patent is the list of claims that describe exactly what the patent is covering, and these can usually be found at the end of the patent or patent application. Claims always start off fairly broad, and then become more specific as you work your way down the list. To fall within the scope of a patent, you only need infringe one of the claims – you do not need to infringe all of them.

The part of the patent between the front page and the list of claims contains detailed information on the invention, together with some examples, and it is usually not necessary to look in any detail at these pages.

Figure 3.5: Example of the front page of a UK patent application. Reproduced with permission. ▶

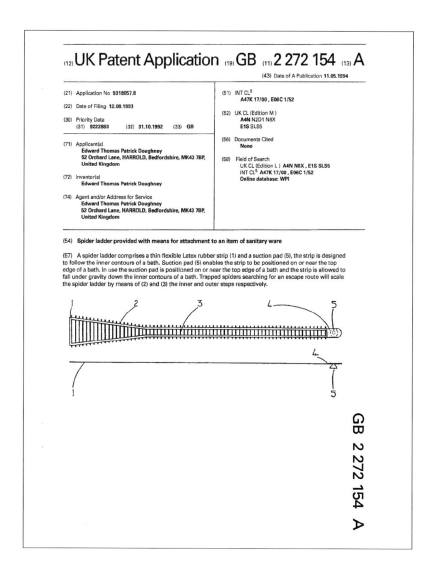

(12) UK Patent Application (19) GB (11) 2 272 154 (13) A

(43) Date of A Publication 11.05.1994

(21) Application No 9318057.8

(22) Date of Filing 12.08.1993

(30) Priority Data
(31) 9222883 (32) 31.10.1992 (33) GB

(71) Applicant(s)
Edward Thomas Patrick Doughney
52 Orchard Lane, HARROLD, Bedfordshire, MK43 7BP,
United Kingdom

(72) Inventor(s)
Edward Thomas Patrick Doughney

(74) Agent and/or Address for Service
Edward Thomas Patrick Doughney
52 Orchard Lane, HARROLD, Bedfordshire, MK43 7BP,
United Kingdom

(51) INT CL^5
A47K 17/00 , E06C 1/52

(52) UK CL (Edition M)
A4N N2D1 N8X
E1S SLS5

(56) Documents Cited
None

(58) Field of Search
UK CL (Edition L) A4N N8X , E1S SLS5
INT CL^5 A47K 17/00 , E06C 1/52
Online database: WPI

(54) Spider ladder provided with means for attachment to an item of sanitary ware

(57) A spider ladder comprises a thin flexible Latex rubber strip (1) and a suction pad (5), the strip is designed to follow the inner contours of a bath. Suction pad (5) enables the strip to be positioned on or near the top edge of a bath. In use the suction pad is positioned on or near the top edge of a bath and the strip is allowed to fall under gravity down the inner contours of a bath. Trapped spiders searching for an escape route will scale the spider ladder by means of (2) and (3) the inner and outer steps respectively.

GB 2 272 154 A

3.4.2 Trade marks

Trade marks are not used to protect inventions, but instead are used to identify products, organisations or services. They are synonymous with brand image, reputation and goodwill, and as such can be incredibly valuable. Like patents, registered trade marks provide a monopoly right, meaning that the owner has the right to prevent others from using that mark in relation to the goods or services for which it is registered. This includes the right to prevent others from using the mark in an internet domain name or as a meta tag on a website. As with other forms of IP, the right to use trade marks can be licensed or assigned to third parties.

The owner of a registered trade mark is entitled to use the ® symbol with their mark. Until registered, or in the case of unregistered marks (see below) the ™ symbol should be used.

Trade marks can be of huge value in the bioscience field, particularly in the pharmaceutical sector. Pharma companies spend vast sums of money developing new products and it is vital for these companies to recoup their costs; they must also make money from the products in order to be able to invest in the development of further new products. However, it also takes a significant amount of time to bring a new product to market, often meaning that there is a relatively short period of patent protection remaining before others can legally manufacture and sell generic versions of the same drug. Pharma companies therefore use trade marks to establish the association of their product name with a particular drug before generic products enter the market. Despite the fact that the drug in generic products is identical to that in the original branded product, consumers will often prefer to buy the branded product; for example, they may choose to buy Viagra® rather than sildenafil citrate.

What can be registered as a trade mark?

A trade mark can be any sign (with some exceptions) that can be represented graphically and used to distinguish the goods or services of one organisation from those of another. Other examples in the bioscience field, besides Viagra®, include names such as Genentech® (the company that manufactures the breast cancer drug Herceptin®), and TaqMan® (real-time PCR probes manufactured by Roche Diagnostics).

Registering a trade mark

There are formal registration procedures that must be followed when applying for a trade mark. These are similar to the procedures that must be followed when filing a patent application. However, the cost of applying for and maintaining a trade mark is much lower than the cost associated with these processes for a patent. The trade mark registration procedure for the UK is detailed on the Intellectual Property Office website (www.ipo.gov.uk).

Trade marks, like patents, are territorial, and they must be registered in each country in which protection is required. Applications for UK trade marks must be filed at the Intellectual Property Office. As with patent applications, there are systems in place that make it easier to file a mark in several different countries. However, unlike patents, there is now such a thing as a European trade mark (known as a Community Trade Mark), which is administered by the Office for the Harmonisation of the Internal Market (OHIM) in Spain.

How long does a registered trade mark last?

Once registered, the mark can remain in force indefinitely provided it is renewed every 10 years – the first ever UK trade mark (the Bass Red Triangle, registered in 1876) is still in use today. A mark may be revoked if it is not used for 5 years or longer.

Unregistered trade marks

Trade marks do not have to be registered to protect the goodwill of a business. Any unregistered sign that is being used to distinguish the products or services of a business is protected by common law. The main advantage of not registering a mark, apart from cost, is that you have more choice over what you can use as a mark. However, there are also certain disadvantages in not registering, including the fact that it is more difficult to enforce an unregistered mark (as the makers of Jif lemon juice found when enforcing their unregistered lemon-shaped bottle mark), and it is therefore recommended that a mark should be registered wherever possible.

Case study 3: Arsenal Football Club

In 2001, the football club Arsenal began High Court proceedings against a street vendor. The vendor, who was selling unofficial Arsenal-branded merchandise, was accused of passing himself off as being connected with the club, and of infringing the club logo trade mark. The court decided that the vendor was not guilty, as passers-by were not confused about the origin of the goods (on the basis that it was clear to consumers that official merchandise would only be sold in shops).

The court also determined that the use of the club logo on the unofficial goods was simply a badge of allegiance to the club, and did not infringe the Arsenal trade mark. Arsenal appealed the decision, and the Court of Appeal ruled that the sale of unofficial branded merchandise does amount to trade mark infringement, a decision that came as a relief to sports clubs across the UK who stood to lose millions in sponsorship if the court found in favour of the street vendor.

3.4.3 Copyright

Copyright is used to protect the expression of an idea, and not the idea itself. It is an unregistered form of IPR that exists automatically once an original work is created, and it is therefore not necessary to formally register copyright. An original work is anything that originates from the author, for example, an artistic work, a literary work, a sound recording, or a film. In the field of science it is more likely to be a publication, a report, an algorithm, a piece of software (e.g. GeneJockey II™ software used for DNA sequence analysis), or a database.

Copyright gives the owner of the original work the right to do certain things in relation to the work, such as copying and issuing to the public.

Generally, the performing of any of these activities by a third party in rela-tion to a copyright work, without the permission of the owner, will be an infringement of the copyright, even where that third party has bought a legit-imate copy of the work, e.g. buying a music CD allows you to listen to the CD, but does not give you the right to copy it or broadcast the tracks on it. Like other forms of IPR, rights to use a copyright work can be transferred to others either through assignment or licensing. It is important to note that unlike patents or trade marks, copyright is not a monopoly right. With monopoly rights, a third party will infringe if they use the invention or mark, even if they have come up with it themselves independently. In the case of copyright, a third party will only infringe a work if they directly copy it. If they create something that is similar or identical to the work completely inde-pendently, then they will not be infringing.

Practicalities of copyright protection

Although there is no need to register copyright, it is usually a good idea to let others know that the work is a copyright work by putting your name, date and the © symbol on the work. In addition, in situations where proof is needed that you were the first to create a work, it may be prudent to post a copy of the work either to yourself or to your lawyer or other professional adviser, preferably using registered post. Once received, the envelope should not be opened, but should be kept as a record that on the date of posting the work had been created by yourself.

Copyright is owned by the person who creates the work or, if that per-son is an employee and the work was created during the normal course of their duties, by their employer. This also applies where a work is commis sioned by a third party – unless the work is specifically assigned to the commissioner then copyright will still belong to the creator.

Copyright usually lasts for the lifetime of the owner plus 70 years, although in some cases the period of protection is shorter, e.g. 25 years for a typographical arrangement (the form and layout of a book or journal).

3.4.4 Database right

As indicated above, databases, which are collections of independent data or works that are arranged in a methodical way, are included in the list of things that can be protected by copyright. However, some databases are also protected by a related and overlapping right, known as database right, which lasts for 15 years. See the UK Government-backed IP website (www.ipo.gov.uk) for further information. Examples of bioscience-related databases include the subscription-based, searchable bibliographic data-bases of publications such as *Web of Science* (wos.mimas.ac.uk) and *CAB Abstracts* (www.cabi.org).

3.4.5   Design law

Design law, which is a particularly complex area of IP law, protects the design of products that are not inventive enough to qualify for patent protection. In the UK the situation is similar to that described earlier for trade marks in that there are two forms of right, registered and unregistered, that protect designs. There is considerable overlap between these two types of right, and these in turn are closely related to copyright. All three forms of IPR can apply in some way to the same article.

Note that in some other countries, including the US and Germany, there also exists a lesser form of patent right called a petty patent, or utility patent, that can be used to protect designs. However, this does not currently exist in the UK.

In certain limited instances in the bioscience field in the UK, a registered design may be used to protect aspects of an invention as an alternative to patenting. This only applies to a physical product, such as a device for making it easier to open Eppendorf tubes, where the entire design is not dictated purely by function; for example, there could be a certain amount of design freedom in how the handle is shaped, making it possible to have a distinctively shaped handle for marketing purposes. In cases such as these, where funding is limited and would therefore not permit the expense of filing a patent application, the use of registered design as an alternative to patenting is particularly useful. The design registration would afford some protection for the invention, although this would not be as strong as the protection afforded by a granted patent.

For further information on design law see www.ipo.gov.uk.

3.4.6   Confidential information

Confidential information, which is also referred to as know-how or trade secrets, may not be classed as IP *per se*; protection and exploitation of an idea as know-how is nonetheless of value, and this concept was discussed in *Section 3.3.2*. Confidential information is any information, owned by an individual or organisation, that is to be kept secret and out of the public domain. Confidential information is protected by the common law of confidence.

As we have already seen, keeping an invention secret may be chosen as a long-term alternative to patent protection. This is only an option where it is not easy to determine the nature of a product through reverse-engineering, e.g. if the product involves binding of an antibody to a membrane in a particular way. If a product is mechanical in nature, then as soon as it is released onto the market it may be possible for an expert to determine precisely how it works. Under these circumstances it would not be possible to protect the product as confidential information.

There are a number of advantages associated with protecting an invention as confidential information. These include the fact that this form of protection will last as long as the information remains secret, which in theory could be in perpetuity, whereas a patent provides a monopoly for 20 years only. In addition, there are none of the high costs associated with patent protection. There are also disadvantages associated with protection as confidential information. The first is that there is always the danger that someone will independently develop and patent exactly the same invention, meaning that your rights to use the invention then become much more limited. The other main disadvantage is that the information is no longer protected if it enters the public domain by whatever route, and it is not possible to restore its confidential nature. In such cases, the only remedies available are to seek an injunction or sue for damages, both of which are expensive options.

From the outset you will need to find out more about how your thoughts and plans relate to the activities of competitors in the marketplace. It can be difficult to undertake this sort of research without revealing something about your own idea, and so you should ensure that your ideas are firmly protected, using the approaches described in this chapter, before embarking on more detailed research of the sort described in the next chapter.

3.5 – Additional resources

Bainbridge DI (2006) *Intellectual Property.* Longman. An in-depth text designed primarily for students of law. Although it probably contains too much detail for bioscience students, it is very readable and is particularly recommended for those who have an interest in finding out more about this complex area.

Newton D *et al.* (1997) *The Inventor's Guide.* Gower, London. A guide to protecting your ideas that is aimed at inventors rather than legal specialists. It is edited by the Patents Information Team at the British Library, and is written in a way that is easy to understand.

Sullivan NF (1995) *Technology Transfer.* Cambridge University Press, Cambridge. This book covers the whole area of technology transfer, and has a very useful chapter on intellectual property in the context of commercialisation of ideas. It has been written specifically for scientists whose work has commercial applications.

UNICO Practical Guides – Confidentiality Agreements and *Material Transfer Agreements*, published by UNICO in association with Anderson & Company. These guides have been put together by and on behalf of UNICO (www.unico.org.uk, a forum for exchange and development of best practice in the University knowledge commercialisation arena). The guides provide general background to each type of agreement, as well as examples of agreements that can be easily used and adapted.

www.bbsrc.ac.uk/biobusiness_guide
BBSRC guide to technology exploitation; section on IP

www.bioindustry.org/
Site of the BioIndustry Association; a publication discussing the patenting of inventions in biology is available on this site

www.cla.co.uk
Copyright Licensing Agency; copyright information relating to photocopying and scanning

www.defra.gov.uk/planth/pvs/
UK Plant Variety Rights Office

ep.espacenet.com
> European Patent Office Espacenet search site

www.european-patent-office.org/
> For information on European Patents

www.google.com/patents
> Patent search site that is a recent addition to the Google family of search facilities; currently only US patents available on this site

www.icann.org/
> International organisation with responsibility for internet address space allocation

www.ipo.gov.uk/
> UK Intellectual Property Office site; for information on patents, trade marks, designs and copyright

www.jpo.go.jp
> Japanese Patent Office

www.uspto.gov
> US Patent and Trade Mark Office

www.wipo.int/
> World Intellectual Property Organisation; for information on PCT patents and international IP

Learning outcomes

Key learning points from this chapter are:

- The importance of thinking about protecting ideas even before they are conceived and then ensuring they are protected through all stages of development and exploitation

- You should never tell a third party your idea, or provide them with samples of proprietary material, without first putting in place a suitable confidentiality agreement, material transfer agreement or other form of agreement with adequate provisions

- There are a number of options for protecting an idea or invention, some of which require formal registration while others do not

- Patents are used to protect inventions; they provide strong protection in the form of a 20 year monopoly, but they are expensive and formal registration is required

- Trade marks are used to protect brand names; they build value in a brand and can be incredibly valuable; trade marking a brand is cheaper than patenting an invention; formal registration is required

- Copyright is used to protect original works, including software and journal articles; no need for formal registration as protection exists automatically

- Database right is similar to copyright and is used to protect databases; no need for formal registration

- Designs are used to protect the design of a product; they can be registered or unregistered and can be an alternative to patenting in a number of limited cases; protection provided by registered design is longer than that provided by a patent; cheaper than patenting

- Information, ideas, and inventions that cannot easily be reverse-engineered can be protected as confidential know-how; extreme care must be taken to protect confidentiality as know-how can no longer be protected if it enters the public domain; no need for formal registration

- The form of protection chosen is determined by several factors including the nature of the invention

- You should think very carefully before publishing an idea as this could prevent the filing of a patent at a later date

- You should consider any publication very carefully even if a patent application has been filed, as publication could prevent the patenting of improvements or new uses of the technology

- The cost of getting it wrong can be considerable – if in doubt seek professional advice

Chapter **4**

Researching Ideas

David Wilkinson & Amanda Selvaratnam

4.1 – Introduction

Having generated an idea and considered how best to protect the IP, you now need to assess and evaluate its novelty. This chapter identifies some of the key tools and techniques available to bioscientists to enable them to research the value and uniqueness of their ideas, specifically, how to:

- use the published scientific literature to establish the uniqueness of the idea (for example, as a tool for finding out if others have had the same or similar ideas)
- assess the competition

Investigating the uniqueness of a potential product or service involves a structured approach and draws upon a wide range of data sources. *Chapter 3* described how patent searches can provide invaluable information about the ideas and work of others. These searches will also provide an indication of the legal status of ideas; that is, the extent to which they are protected. Clearly, published scientific literature in the same subject area as the idea will, like patents, provide a useful indication of what has already been developed. Searches made using a range of web-based sources, including databases, will reveal current developments, emerging trends and the scope of ideas. Where similar ideas already exist, it is important to find out more about the competition through market research and the acquisition of competitive intelligence – the process of gathering and analysing information on the external competitive environment.

All of these approaches are dealt with in this chapter, starting with a consideration of how the published scientific literature may be used when researching ideas.

4.2 – Using published scientific literature to establish novelty

"I think there is a general weakness amongst students – that is the ability to effectively use the literature to inform their own writing and idea development. You receive very little training on it (in undergraduate degrees) and I didn't really know how to make the most out of the literature for the development of my own ideas until I was quite a way through my MSc. I think without this skill, I'd have wasted an awful lot of time researching ideas that were quite 'old hat' and had been done better elsewhere before."

Biology PhD student

It is clearly important to be aware of any articles or items that describe ideas similar to your own. In a literature review a wide range of material, published in journals, textbooks or websites, is interrogated. The results of such a search can be invaluable in that previous successes or failures with similar ideas may be revealed. This information will allow you to rapidly establish the strengths and weaknesses associated with your own ideas.

The summary of different types of literature in *Section 4.2.3* has been ranked, with the most reliable and accurate presented first. Always try to rely on information and data obtained from peer-reviewed sources such as academic books and journals. If data pertinent to your search are found on the internet or in a newspaper, always check the veracity of the information in more scholarly works.

As well as the more traditional published materials, there are other sources of information and intelligence data that you should search when researching the uniqueness or suitability of a business idea. A competently planned and conducted search of available material will help you to think critically both about the material consulted and your ideas as they develop. You should adopt a structured approach to searching and you should consider the following[1,2]:

- the relevance of the material to your own idea(s)
- the purpose(s) of the study or studies reported
- the topicality of the material
- information about data sources (what they were, how many, etc.)
- detail on any substantive results (what they are and what they mean for your idea(s)
- the research methods used
- the breadth of the material
- any exclusions in the material (for example, any areas not investigated?)
- the authority of the material (is it written by an 'expert' in the field?)
- the currency of the material (is it cited by/respected by others?)
- the availability of the material (is it easy to find and access or is its distribution restricted?)

Case study 4.1 illustrates why it is important to use a checklist of this nature when reviewing existing literature and patents (see also *Chapter 3*) during the assessment of the viability of a new product or idea.

4.2.1 Libraries

University and large public libraries offer access to millions of individual sources of information in books, journals, newspapers, videos, DVDs and from online sources. Sifting or searching through all of this material to inform or shape the development of ideas can be enormously demanding. Fortunately, most libraries have specialist staff who will help you with this. A Subject Specialist (or Subject Librarian) should be able to provide useful

Case study 4.1: Establishing novelty

A research project between Dr John Sparrow at the University of York and the Smith & Nephew Group Research Centre on the contractility of skin cells suggested that cell contraction was important for wound healing. However, laboratory tests showed that small specific peptides applied to these cells reduced their contraction, and in animal models actually reduced healing times and scar formation. Apparently contraction forces in the wound, presumably optimised to form a physical barrier as fast as possible, reduce the ability of wounds to heal well. A great commercial product? Well, it never made it.

The decision whether to develop this product further was made based on information from a variety of sources – both publicly available and as commercial expertise within the parent company.

From the scientific literature it was clear that the cells bound skin matrix proteins from which the peptides were derived. The peptides were known to inhibit binding to these proteins. In addition, the peptides were cheap to produce, readily available from biological chemical suppliers, and so available to others.

Market research by the company and their commercial expertise suggested that customers were unlikely to pay much extra for such products. Thus increased income from wound care products enhanced by the peptides was unlikely to generate sufficient extra income to pay for the development, distribution and marketing costs, and also for protecting the product in the marketplace.

Patent searches, etc. by the company revealed that there were a number of patents which it might be difficult to show did not have preference (prior art) to the IP developed by the research. Surprisingly, a major one was a patent for the topical application of snake venom to treat skin ailments and injuries. The scientific literature already contained considerable information suggesting that snake venom proteases would release the active peptides *in situ* from the human skin matrix proteins. This demonstrates just how extensive such IP searches must be and that they can involve a combination of scientific knowledge, the scientific literature and the existing patents.

In the end it was decided not to proceed with this wound care product despite the scientific indications of medicinal and cosmetic value.

advice, including instruction on how best to search for information within the subject field, and training in the use of relevant sources of electronic information.

4.2.2   Keywords and search terms

"I found it useful to follow up on specific material using certain keywords. If particular journals are referenced within the pieces that I find then I'll go and take a look at the recent contents pages of these journals to see what related material they might be publishing."

Research student

When using library catalogues or databases to research ideas you will be expected to enter keywords associated with your subject of interest. For example, you may be interested in the general topic of food preservation.

By entering keywords such as 'food preservation', 'preservation techniques', and 'preservatives', and specifying a search in the title or subject field of publications, a large number of journal articles, reviews, books, etc. on this subject will be identified. To reduce the number of 'hits', you need to focus the search by entering more precise and specific key words. Alternatively, you may consider restricting your search by using search operators such as AND, OR, NOT. For example, you may decide to exclude specific procedures such as hydrostatic preservation techniques from your search. To do this you would insert the NOT command, as follows: 'food preservation' NOT 'hydrostatic preservation'. This will return a list of materials that focus on food preservation but not articles on hydrostatic preservation.

4.2.3 Types of literature

Literature comes in a variety of forms: academic literature; trade/professional literature; and 'grey' literature, a term used to describe material which might not be formally published by commercial publishers, such as working documents and institutional reports. Academic literature, published in journals or textbooks, has a degree of kudos associated with it, particularly material published in journals as described below.

Academic journals

Most academic journals contain material that has been reviewed and refereed by at least two experts (academic peers recognised as having some expert knowledge of the subject matter) prior to the material being published. As a result, material appearing in academic journals can be said to have undergone a quality assurance process. All university libraries subscribe to academic journals, with most of them available electronically. A key benefit of online journals is that they can be interrogated and searched for key words and themes related to your idea.

Professional journals

A professional journal is a hybrid between a typical magazine and an academic journal. These publications are practitioner-focused (written for a particular, usually professional, audience) and they are less formal than academic journals. Nonetheless, they can be useful in that they often provide material related to recent developments in a particular area. An example is the journal *Microbiology Today* published by the Society for General Microbiology.

Textbooks

A good textbook will provide a clear introduction to the subject, offering a structured overview and analysis of an area of relevance to you and your ideas.

Conference presentations and proceedings

Conference presentations and proceedings (collections of papers presented at a given conference) can provide good current accounts of work in progress in your area of interest. Proceedings are often more informal than journal articles or textbooks as they are subjected to less rigorous quality assurance processes. Conference papers are often delivered following the approval of a brief summary of the work by the conference organisers. Nonetheless, conference proceedings are important as they provide an up-to-date indication of exciting developments in the biosciences.

Consultants' reports

These are produced by experts commissioned to write about a given subject area. Such reports are often detailed and can draw upon the expertise of a number of writers collaborating under the banner of a named organisation. An example of such a report is the Oxfordshire Biosciences Cluster sector report written and published by the Oxfordshire Biosciences Network group (www.oxfordshirebioscience.com). This report provides an overview of bioscience developments in and around the Oxfordshire region, with commentary on areas of growth as well as new product developments.

Book reviews

Book reviews are a useful source of information, providing an interesting and critical evaluation of newly published work. Through reading a book review, you can quickly become acquainted with the key issues within a text. Book reviews can be accessed through academic journals, some newspapers, and via online forums/discussion groups.

Newspapers and magazines

Newspapers and magazines usually publish material that is current and attractive to a particular readership. For the more reputable of these publications, material is validated during editing. This, generally, ensures material reported within such sources is reliable. However, some newspaper reports focus upon the 'readability' of material rather than its factual content. Therefore, this source of information should be used with care.

4.2.4 Web-based materials

Part of your research is likely to involve consulting web-based materials – remember that the quality and accuracy of these is variable. Fink[3] lists a number of evaluative questions that should be answered when using web-based materials.

Who supports or funds the site? Are the views of a sponsoring organisation apparent in the material? For example, a web resource funded by a pharmaceutical company may present information or data which present that particular company in a favourable light. It is unlikely to contain material that criticises their work!

What authority do the authors or investigators have? Do the authors of the material have experience in the area? Have they published material elsewhere? Are you or others aware of their work?

Is it clear what the authors did? If the website contains a description of an investigation, do the authors provide a clear indication of what they did and is their approach acceptable? Are data presented clearly? Do the conclusions appear reasonable?

When was the site last updated? Many internet-based resources contain information that is not regularly updated. Is the material you are viewing current and relevant? Look for the 'page/site last updated' line within the website.

4.2.5   Databases

Databases that contain extensive and detailed information about the published literature, are invaluable tools for the researcher keen to explore the uniqueness of their ideas.

University and other large libraries subscribe to a wide range of electronic databases. All of these can be searched quickly for relevant materials using traditional literature search approaches such as keyword/term searching. It is likely that you will be able to download and print entire scientific papers and review articles following locating them via a database search (assuming your library subscribes to the journal).

The ZETOC database

The ZETOC service (zetoc.mimas.ac.uk/about.html) provides access to the British Library's Electronic Table of Contents (ETOC). The database contains details of approximately 20000 current journals and 16000 conference proceedings published each year. With around 20 million journal and conference records, the database covers every imaginable subject in science, technology, medicine, engineering, business, law, finance and the humanities. The database covers the years from 1993 to date and is updated daily. Copies of all of the articles and conference papers listed on the database can be ordered online from the British Library's Document Supply Centre in Yorkshire, or are available electronically if your library subscribes to the journal already. ZETOC is available free of charge to all UK further and higher education institutions.

Consider again the field of food preservation. You may have an idea for a new fruit preservation technique and will therefore wish to find out about existing procedures. If you search the ZETOC database using terms such as 'fruit' and 'preservation', the search will return a handful of records from the database, some of which concentrate upon preservation techniques involving high hydrostatic pressure, whilst others provide a comparison of controlled atmospheric storage against other fruit preservation technologies. The information and analysis provided within these articles might usefully inform the development and progress of your own ideas in relation to fruit preservation.

ISI Web of Science

Web of Science (scientific.thomson.com/products/wos) consists of five databases that contain information gathered from thousands of scholarly journals in all areas of research:

- Science Citation Index Expanded™
- Social Sciences Citation Index®
- Arts & Humanities Citation Index®
- Index Chemicus®
- Current Chemical Reactions®

The three citation databases allow you to determine, for example, how many times an author or a specific article has been cited in the scientific literature. This information can prove invaluable when you must gauge the significance of work in a particular field or from a specific laboratory. Between them the two chemistry databases provide information about a large number of chemical structures, synthetic methods and reaction diagrams. They will lead you to all relevant publications associated with these structures, methods and reactions.

The ISI Web of Science is an extremely powerful database. It allows you to search for terms that may be related to your own business product or service idea. Once these terms have been located within published material captured by Web of Science, you can then search for all material that cites the initial sources you identified in your search. For example, if you wish to continue researching the uniqueness of an idea for hydrostatic food preservation you may feel it will be useful to carry out background research into other applications for hydrostatic preservation. Using the terms 'hydrostatic' and 'preservation' (and restricting your search to finding these terms only in the titles of articles) returns five articles that cover the subject. If you consider any of these articles to be of particular significance, you can use Web of Science to identify later publications that cited the relevant article. This powerful research tool allows you rapidly to explore the work of a large number of individuals who may have similar ideas to your own.

PubMed

PubMed provides access to citations and full-text articles from the biomedical literature (www.pubmed.gov). It is a very large database, but it is specifically focused upon literature in the broad field of biomedicine.

Further hints on searching databases

When you begin to research a topic you are likely to be overwhelmed by the vast amount of information generated by an electronic search. It can help to restrict the amount of material accessed and a simple way to do this is to specify a time interval for the search. For example, if you are interested in the very latest developments in a particular subject, then it would make sense to restrict your search period to material published in this field during the last year. To focus your research even further, you may wish to consult only those sources (journals) that are the most widely respected in a particular field. A tool to help you do this is the *Journal Citation Report* service (this will usually be provided as part of your library's access to Web of Science or Web of Knowledge) which provides an indication of the most popular journals in a wide range of discipline areas or sectors. In essence, the most important journals are deemed to be those that receive the most citations for the articles they publish. This ranking is called the impact factor for the journal. In the field of biochemistry and molecular biology, two important journals are the *Annual Review of Biochemistry* (impact factor of 33.45) and *Cell* (impact factor of 29.43). This means that, on average, articles in these two journals are cited by about 30 other scientific publications per annum.

4.3 — Competitive intelligence

Once an idea for a product or service has been formulated and researched, using the scientific literature and databases, it is essential that an analysis of the potential market for the innovation should be undertaken. There is no point in developing a new product or service if there is no demonstrable market need or significant benefit when compared to existing products or services, or if there are too many competitors selling similar or alternative products/services. The importance of market analysis during the preparation of a business plan is stressed in *Chapter 7* and a key component of that analysis is the acquisition of 'competitive intelligence' (CI). CI can be defined as the process by which information about products, customers and competitors is gathered and used for business planning. The term is occasionally used synonymously with 'competitor intelligence' (or competitor analysis) but CI involves much more than analysis of competitor behaviour; it is about gathering data to inform the development of a *strategy* to ensure that your business is more competitive and successful than the businesses of existing and potential competitors.

CI gathering should be a continual process and should play an integral role within any business, no matter how large or small. Successful CI gathering can enable a company to minimise threats and maximise opportunities by providing an awareness of the market drivers and the competitive environment. There are numerous examples of individuals or companies who have significantly increased their income using information gained through successful CI gathering activities. It should be stressed that while, from time to time, individuals may be tempted to behave in an unprincipled and underhand manner during the acquisition of CI, there is no reason why effective CI gathering need be carried out in anything other than an entirely legal and ethical (see *Chapter 10*) framework.

4.3.1 CI professionals

Clearly the acquisition of CI will be essential for the development of a biotech firm's business strategy. The scientists and academics associated with the company will acquire a great deal of CI during their day-to-day lives and, in the remainder of this chapter, advice is provided about how these 'amateurs' can optimize acquisition and exploitation of CI. However, when it becomes clear that the new business must undertake a more detailed CI gathering exercise, the company should engage the services of CI professionals. For a bioscience start-up, these individuals will have an intimate knowledge of the biotechnology industry with a wide network of contacts. They will have the ability to collect intelligence from a wide range of sources, distinguish relevant information from irrelevant information, analyse large amounts of data, assess developments within the industry, and achieve a synthesis of the knowledge acquired from these disparate sources. With their help it should be possible to identify the most appropriate actions for the biotech firm (see *Case study 4.2*). Further information about CI professionals can be obtained from the Society for Competitive Intelligence Professionals at www.scip.org.

4.3.2 Information gathering

At the start of this chapter the structured approaches you should adopt, when undertaking literature searches to help to develop and focus your business ideas, were described. The acquisition of CI requires a similarly structured approach. There are essentially two broad forms of data you should collect – primary information and secondary information.

Primary information

Primary information tends to be unpublished information and, while it is often more difficult to obtain and verify, can frequently provide the most

Case study 4.2: Competitive intelligence
Using CI to attack and defend a position

A CI team at a biotechnology company discovered that a competitor had a serious manufacturing problem that was going to result in the temporary withdrawal of a product. This information was reported to senior management who developed a marketing strategy to take to the competitor's customers and fill the gap. The company earned millions of dollars simply because they knew that a competitor was about to withdraw a product.

Of course, there are also many examples of how companies have lost significant income by not undertaking sufficient market research and thus failing to acquire competitive intelligence. One such example involved Coca Cola. In the mid-1980s, Coca Cola faced some serious competition from its main rival, Pepsi. Instead of developing new marketing and promotion strategies, Coca Cola decided to introduce the world to 'New Coke'. It was a commercial disaster as they failed to undertake sufficient market research to determine whether customers would actually prefer the new taste. In the end New Coke sales plummeted and the company had to revert to selling Old Coke, having wasted significant resources developing and marketing a new brand that no-one wanted.

It is always important to assess not only the competition but also to understand your own products, services and customers.

useful intelligence. You can obtain primary information from the following sources:

- interviews with experts in the field, including customers and suppliers
- marketplace surveys
- interviews with industry analysts
- trade associations
- conferences and exhibitions
- unpublished documents
- recruitment agencies
- internal sources

Internal sources include the knowledge and data contained within colleagues' minds, filing cabinets and computer hard drives and so, of all the sources of information, these are probably the most difficult to access.

Secondary information
Secondary information is information that has already been published and is therefore immediately available. When starting to research a market, a good place to begin is by looking closely at another business that is doing something close to what you intend; potentially a future competitor. Visiting a competitor's website can often yield a huge amount of relevant information or competitor intelligence. A detailed list of sources of secondary information is provided at the end of this chapter. Important among these are:

- websites of competitors
- websites and publications of potential investors
- databases with links to academic and other publications
- industry and government reports, and other information, including the material published by UK Trade and Investment (see www.uktrade invest.gov.uk)
- directories
- statistical sources
- newspapers and magazines
- reputable trade publications
- recruitment adverts
- libraries, including specialist business library services such as www.bl.uk/knowledgetransfer
- patents
- regulatory requirements

We shall now consider a number of sources of primary and/or secondary information in greater detail.

Marketplace surveys. As an idea for a product or service is developed, it is essential to understand the market in which it will be exploited or sold. It is not always obvious where or who the market will be. For example, a scientist developing a compound to reduce blood cholesterol levels may wish to approach pharmaceutical companies to ascertain their interest. However, there may be other options open to the scientist including, for example, approaches to health food or food processing companies. The first priority should therefore be to identify the market or markets with the potential to exploit the product.

Once the target market has been identified, an understanding of the size of the market, and the competition within the market, is needed. Bioscience products frequently take a very long time to develop. The market will almost invariably change and develop during this period. It is therefore important to keep a watchful eye on competitor activities, and identify new and emerging competitors. This activity can be facilitated in a number of ways, for example, by recruiting a scientific advisory board with expertise in the relevant fields, networking at conferences, or working with specialist groups.

Investors and finance. One of the most vital tasks for the leader of any start-up business is to ensure that the resources that the firm needs to operate are available. During the acquisition of CI it should therefore be established which venture capitalists are investing in the relevant area of technology and the organizations that might be interested in acquiring these technologies. These data can help establish the value of a company which can help in situations that involve negotiation, such as licensing. In addition, this aspect of CI can be used to examine potential merger candidates or joint-venture partners who might provide alternative sources of funding (see also *Chapters 6* and *8*).

Academic and trade publications and conferences. A thorough examination of the research and market environments can be conducted through the analysis of academic and trade journals. It should be noted, in particular, that information about very recent developments can be gleaned from the proceedings of scientific and trade conferences. Trade sources will provide essential information about how a newly introduced product or service will be marketed and distributed. This information will enable a new business to develop an efficient and targeted strategy for the marketing and distribution of its own product(s) or service(s).

In addition to the essential knowledge that publications, conferences and patents (see below) will provide regarding competitor activities, information from these sources can also be valuable for a number of additional reasons including the following:

- It can make entrepreneurs aware of applications for newly developed technologies. It is essential that entrepreneurs in the biosciences keep up to date with new developments and remain fully aware of the technologies being used or developed by their competitors. This approach will provide an invaluable insight into the opportunities and problems associated with these new technologies.
- It may also help identify opportunities to collaborate with other organisations that are able to bring key skills or experience to the development of a new product or technology. For example, many bioscience start-ups lack the expertise to take a product to market. These fledgling companies may wish to consider the option of collaborating or partnering with other organisations. By monitoring new technologies entering the market and in development, potential partnerships can be identified.

Recruitment adverts and agencies. Analysis of recruitment adverts can provide important insights into competitors' staffing strategies. Likewise, recruitment agencies, while keeping their client information confidential, may also be good sources of information about industry skill trends and the strategies of non-client firms. This kind of information can allow a company to determine the type of people it needs to attract to succeed in a market niche, and what it will take to attract and retain them.

Patents. Patents are a source of information whose value is often overlooked. They are used to protect corporate IP (see *Chapter 3*) but also offer a wealth of scientific and technical information. Virtually every country grants patents and it is believed that between 70 and 90% of information in patents is never published elsewhere. This means that patent databases are novel and extremely useful sources of knowledge and data. Ideally a comprehensive patent literature search should be conducted at least annually. This can either be contracted out to an organisation or done in-house. Sources of information on patents are provided in *Chapter 3*.

Regulatory issues. It is essential to be aware of current regulatory requirements and to identify issues that might affect the approval of a product or the way it is marketed (see *Chapter 9* for a detailed consideration of regulatory issues). The acquisition of CI should therefore involve a survey of the regulatory agencies. A new business should carefully consider the regulatory environment and anticipate changes that may profoundly affect its activities and the activities of its competitors.

The information gathered from the above sources should be used to compile a comprehensive database that should contain information about potential customers, the political and regulatory environment, sources of funding and detailed profiles of the key players. The latter should include information about competitors':

- marketing and branding strategies
- sustainable competitive advantages
- R & D and other operations
- key personnel
- organisational structures
- major customers
- intentions regarding customer groups

The profiles should also contain information on how the investment and financial community perceive the branded value of competitors' products and services.

4.3.3 Analysing the intelligence

It is likely that information will have to be obtained from several sources to ensure that it will answer effectively the key questions asked during the acquisition of CI. For example, you may wish to know how much a competitor will spend next year marketing a new drug. Of course, it is very unlikely that this information will be available in the public domain. However it may be possible to find out by combining information from a number of sources. Thus, the competitor's annual report may refer to the expected revenues for the new drug while another information source, such as a market research report, may reveal that the competitor is planning to spend 10% of revenues on marketing. Combining these two pieces of information will give an approximate marketing budget for the drug.

It is essential that the information gathered is validated. Some of it may be out of date, misleading, incomplete, inaccurate, or simply wrong. The validity of the information can be assessed using a ranking system based on factors such as whether:

- the source of information has been reliable in the past
- the information can be corroborated by other, independent sources
- the information makes sense and fits with what is already known

Once validated, the data should be presented in a form that will allow the key questions to be answered readily. A table that summarises the competition and provides an indication of the percentage market share of other businesses is a good starting point. Such a table can be augmented with the comparative strengths and weaknesses of the competitors relative to those of the start-up business.

Information can also be analysed by well-established techniques such as PEST (Political, Economic, Social, Technological) or SWOT (Strengths, Weaknesses, Opportunities, Threats) analyses. In PEST analysis the market, including competitors, is assessed from the perspective of your business or proposition. It involves an analysis of external factors. SWOT analysis, on the other hand, involves an assessment of your own business proposition, or the proposition of one or more of your competitors. It tends to involve an analysis of both internal and external factors. It is useful to undertake PEST analysis prior to SWOT because PEST often helps identify SWOT factors. Inevitably there is some overlap between PEST and SWOT analyses.

PEST analysis. PEST analysis can be undertaken most effectively by constructing a grid (*Figure 4.1*). For each heading (political, economic, social, technological) an entrepreneur should think of every external factor that could possibly have an impact on the business, and the inter-relationships between these factors. The more connections there are between any one factor and other factors, the greater the impact that factor may have on the business and the greater the need for it to be considered when devising a business strategy. The analysis can be further extended by giving a score to each of the items in each section. This can be particularly useful, as a comparative exercise, if more than one market or opportunity is being analysed.

SWOT analysis. SWOT analysis is a strategic planning tool used to evaluate the strengths, weaknesses, opportunities, and threats involved in a business venture. The strengths and weaknesses apply to internal factors, and the opportunities and threats usually relate more to external issues. SWOT analysis enables an in-depth analysis of your own organisation and a direct

Political	**Economic**
• ecological/environmental issues	• home economy situation
• current legislation in the home market	• home economy trends
• future legislation	• overseas economies and trends
• European/international legislation	• general taxation issues
• regulatory bodies and processes	• taxation specific to product/services
• government policies	• seasonality/weather issues
• government term and change	• market and trade cycles
• trading policies	• specific industry factors
• funding, grants and initiatives	• market routes and distribution trends
• home market lobbying/pressure groups	• customer/end-user drivers
• international pressure groups	• interest and exchange rates
Social	**Technological**
• lifestyle trends	• competing technology development
• demographics	• research funding
• consumer attitudes and opinions	• associated/dependent technologies
• media views	• replacement
• law changes affecting social factors	• technology/solutions
• brand, company, technology image	• maturity of technology
• consumer buying patterns	• manufacturing maturity and capacity
• fashion and role models	• information and communications
• major events and influences	• consumer buying mechanisms/technology
• buying access and trends	• technology legislation
• ethnic/religious factors	• innovation potential
• advertising and publicity	• technology access, licensing, patents
	• intellectual property issues

Figure 4.1: PEST analysis. ▲

comparison with the strengths, weaknesses, opportunities and threats for competitors. As with PEST analyses, it is useful to construct a grid, in this case containing all known strengths, weaknesses, opportunities and threats (*Figure 4.2*). This will allow you rapidly to assess the significance of each of these factors.

PEST and SWOT analyses can prove particularly valuable when developing an entry strategy for a product or service. For example, if a new cholesterol-reducing nutraceutical (a health-promoting food) is being developed using transgenic crop technology, then a PEST analysis would,

Strengths (internal)	Weaknesses (internal)
• advantages of proposition	• disadvantages of proposition
• capabilities	• gaps in capabilities
• competitive advantages	• lack of competitive strength
• USPs (unique selling points)	• reputation, presence and reach
• resources, assets, people	• financials
• experience, knowledge, data	• own known vulnerabilities
• financial reserves, likely returns	• timescales, deadlines and pressures
• marketing – reach, distribution, awareness	• cash flow, start-up cash-drain
• innovative aspects	• continuity, supply chain robustness
• location and geographical	• effects on core activities, distraction
• price, value, qualify	• reliability of data, plan predictability
• accreditations, qualifications, certifications	• morale, commitment, leadership
• processes, systems, IT communications	• accreditations, etc
• cultural, attitudinal, behavioural	• processes and systems, etc
• management cover, succession	• management cover, succession
Opportunities (largely external)	**Threats (largely external)**
• market developments	• political effects
• competitors' vulnerabilities	• legislative effects
• industry or lifestyle trends	• environmental effects
• technology development and innovation	• IT developments
• global influences	• competitor intentions – various
• new markets, vertical, horizontal	• market demand
• niche target markets	• new technologies, services, ideas
• geographical, export, import	• vital contracts and partners
• new USPs	• sustaining internal capabilities
• tactics – surprise, major contracts, etc	• obstacles faced
• business and product development	• insurmountable weaknesses
• information and research	• loss of key staff
• partnerships, agencies, distribution	• sustainable financial backing
• volumes, production, economies	• economy – home, abroad
• seasonal, weather, fashion influences	• seasonality, weather effects

Figure 4.2: SWOT analysis. ▲

amongst other things, help to identify both the countries in greatest need of such a product and the political implications of growing and selling the transgenic product in certain countries.

It is probably also apparent to you that PEST and SWOT analyses can be used in all sorts of settings. For example, you may wish to use these techniques when you are trying to decide whether to take a year out in industry, rent a flat, buy a car, or become a bioentrepreneur!

In summary, a detailed consideration of CI data using PEST, SWOT or other analytical procedures is an essential part of researching ideas. In the absence of a detailed analysis the information is of little or no value and the painstaking effort involved in data acquisition becomes no more than a fruitless exercise. The results of the analysis will be of crucial importance as they will inform strategic decisions and allocation of resources during:

- product/service development
- staff and budget allocations
- the development of sales and marketing strategies
- identification of strategic alliance partners

All this information and analysis should lead you to a clear understanding of your competitive position. However, to be really effective you need to adopt a strategy that will ensure your competitive advantages are sustainable. A good example of this would be the filing of a defensible patent that prevents or delays entry of competitors to the market (see *Chapter 3*). You should also be aware that as soon as you and your organisation start to ask questions, write patents, or publish information, your activities will become visible to your competitors. It is therefore very important to carefully manage your 'footprint', especially during the start-up phase of a bio-business. The use of NDAs or other contractual clauses will be of particular value during this sensitive period of idea/product development (see *Chapter 3*).

Now that your ideas are protected and thoroughly researched, they must be communicated to potential investors and customers. In the following chapter, a range of oral and written methods of conveying your thoughts and ideas to individuals, with and without a scientific background, will be described.

4.4 – References

1. **Birmingham P** (2000) Reviewing the literature. In: *Researcher's Toolkit: the complete guide to practitioner research*, Wilkinson D (ed.). Routledge, London.
2. **Bruce C** (2001) Interpreting the scope of their literature reviews: significant differences in research students' concerns. *New Library World,* **102**: 158–165.
3. **Fink A** (2005) *Conducting Research Literature Reviews: from the internet to paper.* Sage, London.

4.5 – Additional resources

The following list is by no means exhaustive but should give an indication of the range of sources of information available for assessing the value of an idea. Due to the need for up-to-date information regarding CI many useful sources are internet-based. Books and other printed materials go out of date quickly and are therefore of less value than sources of information submitted online on a regular basis.

www.google.com
> Google is one of the very best search engines. Within Google there is a sub-site called, 'Google Answers', which allows you to submit questions to screened researchers who will provide answers for as little as £1.50. Before you do this you should first use the main part of the site which may provide you with the required answers free of charge.

Company information sources

www.corporateinformation.com
> Global corporate information on the leading companies in over 55 countries. Subscription to this database provides access to company reports.

www.dnb.com
> Dun and Bradstreet provides access to information on 17 million companies worldwide.

www.icc.co.uk
> ICC is a UK company information database specialising in providing in-depth business information on all UK and Irish companies, directors and shareholders. Information is provided through a wide range of flexible online 'business' and 'integration' services.

International business information sources

www.direct.gov.uk/Homepage/fs/en
> UK Government page with links to all public services.

www.erc-world.com/
> Market reports covering the Food, Drinks, Tobacco, Ophthalmics, Personal Care and Hygiene, OTC pharmaceuticals, and Household Goods sectors in 80 countries.

www.nationmaster.com
> NationMaster is a vast compilation of data from such sources as the CIA World Factbook, UN, and OECD. Demographic maps and graphs based on all kinds of statistics can be generated.

www.the-chiefexecutive.com/
> An international journal covering the CEO's evolving mission and responsibilities through analysis, case studies and interviews.

www.tsnn.com/
> Provides access to conference and trade show web sites for a wide range of industries and sectors, worldwide.

www.uktradeinvest.gov.uk
> An excellent source of international market information.

www.un.org
> United Nations website providing news, statistics and political information.

www.worldchambers.com
> Links to over 12 000 registered Chambers of Commerce and Industry, and Boards of Trade.

www.wto.org
> World trade organisation providing information and statistics.

Market research sites

www.bccresearch.com
> BCC Research provides market research reports, newsletters, and information about conferences. BCC's information products explore major market, economic, scientific, and

technological developments for business leaders in industrial, pharmaceutical, and high technology organisations.

www.datamonitor.com/
Datamonitor produces reports on international business and consumer markets.

www.dialog.com/
Dialog is probably the most comprehensive source of online market research databases in the scientific and biomedical fields. Subscription service with a range of payment options including pay-as-you-go. Dialog includes 'TableBase' – a database comprising tabulated information on: market share, market size, capacity, production, imports, exports, sales, product and brand rankings, forecasts, healthcare statistics and demographics.

www.euromonitor.com
Euromonitor offers quality international market intelligence on industries, countries and consumers.

www.freedoniagroup.com
Freedonia provides a reliable assessment of industry and includes product and market forecasts, industry trends, threats and opportunities, competitive strategies, market share determinations and company profiles.

www.frost.com/
Frost and Sullivan provide free access to press releases, seminars, details of conferences. Online market research report summaries can be viewed for free and purchased online. The database covers all sectors.

www.galegroup.com/
Business and industry site providing information on market size, market share, shipments, users, test marketing, company and industry forecasts, trends and demographics. All major industries are covered in business and industry.

www.keynote.co.uk/
Keynote Reports provide free executive summaries of market research reports. Most reports can be purchased from other sites such as Dialog (www.dialog.com/).

www.marketresearch.com/
Provides access to over 50 000 market research reports from 350 content providers via a pay-as-you-go service.

www.marketsearch-dir.com/
The Market Search Directory offers 20 000 reports from 700 suppliers.

www.mindbranch.com/
Provides access to a comprehensive online catalogue of over 90 000 market research reports, company profiles, newsletters and subscription services provided by more than 400 top publishers and consultancies.

www.researchandmarkets.com/
Research and Markets provides pay-as-you-go access to nearly 56 000 international market research reports.

www.scip.org
The Society for Competitive Intelligence Professionals; provides further information about CI professionals.

www.the-list.co.uk/
The List is a searchable pay-as-you-go database of thousands of executive summaries of predominantly UK and European market research reports, which you can browse, free of charge and buy the complete report online if you wish.

News and media sources

www.publist.com/
A database of over 150 000 magazines, journals, newsletters, and other periodicals.

www.spgmedia.com
> Provides links to controlled circulation magazines, internet reference portals and business conferences and summits.

www.thepaperboy.com/
> Provides links to 1000s of newspapers around the world.

Pharmaceutical information

www.drugdevelopment-technology.com/
> Offers readers insights into new drugs currently in development.

www.ims-global.com/
> International coverage of the healthcare and pharmaceutical markets.

www.pharmaceutical-technology.com/
> Reports on large-scale manufacturing of drugs after the approval stage. These two websites cover everything from compound screening to delivery of a new drug to retail, with detailed editorials, industry news, events listings and more.

www.worldpharmaceuticals.net/
> Information source for executives and researchers in the top 50 pharmaceutical companies worldwide with a unique biannual review of trends and developments across the industry.

Statistical information

www.census.gov/main/www/stat_int.html
> Provides access to national statistical agencies around the world.

UK information sources

www.taforum.org
> Provides links and access to information about UK trade associations and business sectors.

Patent and trademark sources

www.derwent.com
> Derwent are a patent supplier with information from 40 patent offices worldwide. Available through Dialog (www.dialog.com/).

gb.espacenet.com
> Provides access to European patent databases.

www.micropat.com
> Micropatent trademark and patent information supplier providing online and hard copy patents via a pay-as-you-go service.

www.uspto.gov
> The US Patent Office offering access to full US patents online.

Learning outcomes

Key learning points from this chapter are:

- How published literature, web-based materials and databases can be used to establish the novelty of your ideas

- The importance of competitive intelligence not only for gathering information about the market and competitors but also as an essential analytical procedure during the formulation of strategy

- The wide range of sources of primary and secondary information

- The importance of validating the quality and accuracy of all information gathered

- The value of structured analytical procedures like PEST and SWOT analyses

- The importance of carefully managing a bio-business' 'footprint' during the start-up phase

Chapter **5**

Communicating Ideas

Samantha Aspinall & David Wilkinson

5.1 – Introduction

"As a researcher, I've got lots of new and interesting ideas. I keep most of these to myself – I don't think it's a good idea to share them with others. I'm much happier by myself thanks! I avoid working with others and communicating with them at every opportunity."

Microbiology research student

Unfortunately for the student above, ideas cannot truly be developed unless they are refined, tailored and ultimately communicated to others. The excitement of having a new idea that we have worked hard to progress and develop, can make the best of us blinkered and restrictive in terms of the objective assessment of its worth and value. Communication is therefore crucial for the development of our thoughts and ideas. We all need to communicate with others to develop, explain and defend our work, and to seek information or data based on the work of others. Effective communication can help us to focus our ideas, re-shape them and make them better. As a scientist and entrepreneur you must communicate your work and ideas to other scientists at scientific meetings or when seeking funding from research councils or other interested parties. In this chapter, hints and tips are provided about how you should present yourself and your ideas using oral, written and poster communications. The challenges you are likely to face in communicating scientific concepts to those without a scientific background are also considered. Finally, the benefits of networking are discussed along with an explanation of why it is increasingly important that bioscientists should interact effectively with a wide range of individuals from all walks of life.

5.2 – Presenting yourself and your work

When you want to communicate an idea, understanding how to effectively present yourself to your audience can be the key to success. In this regard, it is important to be aware of how other people perceive and understand you and your ideas. A popular technique that helps with this is 'Johari's window'. This widely used interpersonal communication tool was created in 1955 by Luft and Ingham[1].

Johari's window can be used to help you think clearly about how you want to promote and communicate your ideas, and feedback from others will ensure that you develop an appropriate and effective communication strategy. You might like to try writing down:

- 4 or 5 words in the 'Arena' box that describe elements of your personality that you are happy to share with others

	I know this about myself	I do not know this about myself
Other people know this about me	**Arena** These are the things I know about myself that I am happy for others to know. They are my values, my source of motivation, my way of life.	**Blind spot** These are things that other people know about me but that I am unaware of. They are attributes that others can see; feedback regarding these characteristics will help me learn about myself.
Other people do not know this about me	**Hidden agenda** These are the things that I prefer to hide from people.	**Unknown** Neither I nor others are yet aware of these features. There may be considerable potential in exploring this unknown territory.

Figure 5.1: Johari's window (based on Luft & Ingham[1]). ▲

- 4 or 5 words or statements in the 'Blind spot' box that summarise how you think other people see you; consult others about this and find out if your perceptions are accurate
- anything you know about yourself that you do not want other people to know at this time can go in the 'Hidden agenda' part of the window; think carefully about items in this category; you may decide that the 'risk' involved in sharing thoughts and ideas with others is entirely justified
- 2 or 3 words or statements in the 'Unknown' part of the window that describe what you would like to become in the near future; this is your potential

Another crucial aspect of presenting yourself and your ideas effectively is your 'image'. The way a person is dressed and appears can often make a strong impression on the onlooker. In other words, most people make snap judgements, correctly or incorrectly, based on how people look. An archetypal image of an eminent scientist is often one of a man, perhaps dishevelled in appearance and maybe with some slight eccentricity in his clothing, for example, wearing odd socks. This is probably not an image that you seek to cultivate at this stage in your career! So what image would you like to portray? Ask yourself – who am I addressing?

Presenting your work in a tutorial will be very different from presenting an idea to someone who might provide funding and help you achieve a business goal or develop your ideas in some other way. You may also be asked to make a presentation in a situation where there is an element of competition. How will this affect the image you wish to convey? For example, you might enter the Biotechnology and Biological Sciences Research Council (BBSRC) 'YES' competition (www.biotechnologyyes.co.uk) where entrants must work together as a team to develop a business plan for an imaginary biotech start-up company. You and your team-mates may decide to introduce a 'dress code' for your team. If marks are given for the apparent cohesiveness of the team, a uniform image might be in your favour.

You need to feel comfortable and confident about how you present yourself and communicate your subject. It is important to feel as relaxed as possible and one way to do this is to practise slow, measured breathing. Standing with a good posture and shoulders back allows you to breathe freely and keeps your tone of voice steady and calm, making you sound more confident. 'Pitching' an idea that you feel passionately about can make you feel very nervous as you may think that you have a lot to lose if your presentation is not successful. Try to relax to allow yourself to give a calm and well-paced presentation. Remember, if you can convey your dedication and enthusiasm for your work and ideas the audience can't fail to be impressed!

5.3 – Spoken communication

"It's important to discuss ideas with others who know a bit about your vague area of study and can offer their views and interpretation on what you are doing. Bouncing ideas around, and informal discussions are often the most valuable part of any day. You can be really wrestling with some thorny issue, theory or position and by discussing it with others, the mists can clear a bit and you can stand back from your ideas and objectively assess them."

Researcher, University of Edinburgh

Communication is a multi-faceted word that covers almost any form of interaction with another person. This ranges from: a conversation on the bus; buying something; writing a letter or a job interview. For a lot of our daily interactions, we rely on our abilities to speak, listen and read non-verbal signals. It is a mostly sub-conscious process. For example, if you are going to fill up your car with petrol and pay for it, you will not, as a rule, plan how you are going to address the cashier or think about the words you are going to use. Possibly the only part of the communication that you will need to 'plan'

Simon Singh
Brief interview with the writer, broadcaster, and communicator of science

Why do you think it is important to communicate science?

"I think there are two reasons:

Just curiosity: some research is so fundamental and doesn't necessarily have direct benefit to the man and woman in the street, but it is just interesting. It is just wonderful what humans do.

We study the universe around us and we learn some pretty amazing things and it is only right that we pass that on to the public who may become similarly inspired in the future if they are young, they may even want to be scientists in the future.

Much of the research that goes on these days directly affects the public. Whether it is: GM food; stem cells; global warming; issues around nuclear power – the public have a right to be informed. They also have a voice in these issues and unless they are well informed then society may well end up making bad decisions about global warming, GM foods and so on."

Can you give me an example of some really difficult concept that you have had to communicate?

"It is horses for courses. I can talk about cosmology to 10 year olds or I can talk about cosmology to undergraduate physicists. At every stage you just tailor what you are saying or what you are writing and that's it. Just putting yourself in the position of reader and listener and cutting it down so that they will understand it.

News programmes have a list of words that they cannot use without explanation and there was a time when you couldn't use the word genetic without explaining what it meant. Then about 20 years ago, ITN said we can use the word genetic and we can assume that everyone knows what it means.

Similarly with global warming, 20 years ago it was quite a complicated picture for people. Now most people know that it means increase in carbon emissions from burning fossil fuels."

Have you ever had to communicate an idea to get funding for something? For example if you were putting together a brief for a series to be funded?

"You always tailor what you are saying for your audience. If I am doing a live interview the first question I ask is how long have I got? If I know it is 5 minutes, then I know the main points I have to make. If I know I have half an hour then I know how to pace it and deliver it. If I am asked to do a radio show I will find out the listenership. If I am asked to talk in a school I will ask: "is it a maths group?" and "are they there voluntarily?" I always like to know my audience."

Do you think universities have an obligation to help young scientists to effectively communicate what they are doing?

"I think it is great that scientists communicate with the public but there is no reason why there should be compulsion. If somebody is a great research scientist who is a terrible speaker then frankly, they should stay in the lab.

If you have a talent for writing or talking and you enjoy it then I think it is fantastically important that universities celebrate that."

Did you become interested in communicating science through someone you admired in this field?

'I grew up in the 70s watching TV boffins like James Burke. In fact I dedicated my last book to them. I enjoyed it, I was good at it and I had jobs which put me in the front line of persuading young people to enjoy science and it grew out of that.

Really, what it is about is – if you are going to do science communication, do it well."

will be to give him or her the number of the petrol pump you used. This type of communication usually works well for us as we go about our daily lives.

However, when asked to communicate in a more formal setting, an individual will often start to think in detail about the forthcoming presentation. This can make them feel anxious and they may lack confidence in their ability to communicate effectively. In this section these issues are addressed to try to help you understand how to present yourself and your work with confidence, be it formal oral communications or the less formal 'elevator pitch'.

5.3.1   What makes a successful oral presentation?

Many students feel very nervous about giving presentations – this is natural as very few people look forward to speaking in public. Whether you are presenting to peers and your tutor, or a panel of people who you are trying to persuade to support your idea, the principles are the same.

Preparation
- Find out how much time you have for your presentation and make sure you keep to that time – going over your allotted time invariably creates a bad impression (you might even find that the chairperson intervenes to prevent you talking beyond the scheduled end of your presentation and you may be unable to deliver the main message from your talk).
- Establish the size and location of the room – if possible visit the room first to ensure that everyone in the audience will be able to see and hear you. If there is a lectern, do not stand behind it. You may be tempted to use the lectern to create a barrier between you and the formidable mass of people, but your physical presence will be more powerful if you stand directly in front of the audience.
- Check on the availability of white boards, projection facilities, computers, etc. – nowadays, most presentations are made using PowerPoint.
- Plan your presentation: ensure you are clear about the purpose, theme and main points. You should feel passionate about your idea and keen to share it, but the preparation must be thorough to help others follow your line of thought.

Consider the following example of a presentation you might have to make:

- The aim is to persuade members of the audience to fund your proposal.
- The theme of your presentation is that the lack of clean fresh water is threatening the well-being of a growing number of people on the planet – the key point is that there is a need to identify new sources of clean water in developing countries under increasing pressure from rapid population growth.

- Your solution is to make available a cheap water testing kit throughout the developing countries.

The structure of your presentation can now be considered in some detail.

Content

Introduction. Listeners need to be led to the point where you will try to sell your idea. Introduce yourself then try to capture their interest immediately. Don't assume that they have a great deal of background knowledge. Make your initial comments strong with some context. For example, you could start with a powerful statistic: '*1.1 billion people have no access to clean water...*'.

Your introduction will allow people to decide whether or not they want to carry on listening to you. You need to outline the topic by providing a small amount of background information. Introduce your idea, put it into context, and let your audience know you mean business!

The main body. This is the part where you sell your idea. Outline data/research findings that will help strengthen the need for your business proposal, but do not overwhelm your audience. Make sure the presentation is user-friendly: convert data into charts; provide brief summaries of elaborate points; keep to one topic per slide; highlight ideas using bullet points; use examples and analogies to illustrate complex issues, for example, Richard Dawkins[2] describes how the DNA of the human genome can be regarded 'as a set of instructions for how to make a body'.

Think carefully about the number of slides you will use. A 'rule of thumb' is to allow 2–3 minutes for each slide and to make sure you leave time for questions from the audience at the end of the presentation.

Keep animation on your slides to a minimum. Showing the audience that you have learnt how to use clever PowerPoint animation techniques is not a guarantee of an effective presentation; in fact, excessive use of this approach could easily distract the audience who may then miss your main message.

> "The more elaborate our means of communication, the less we communicate."
>
> Joseph Priestley, 1733–1804

Always be ready for the 'worst case' scenario. For a particularly important presentation, if you are using PowerPoint or another electronic audiovisual aid, you may wish to have some form of back-up plan (e.g. transparencies for an overhead projector) in the event of a malfunction, or some other form of complication with the projection facilities.

Conclusion. This needs to be concise, to the point and should reiterate your aim(s). Do not, under any circumstances, end your presentation on a weak note. People often spend a great deal of preparation time on the introduction and main body of the presentation, imagining that the ending will somehow follow on, naturally. Think carefully about how you wish to conclude your talk and signal the end of the presentation by making a remark such as '…in conclusion I would like to say…'. Finish strongly, delivering a clear take-home message and, where appropriate, leave your audience with a firm idea of what you expect of them.

Give your presentation at a measured pace and avoid jargon – this is crucial if you are to get your message across. It is easier for your audience to understand your message if you communicate in short, simple sentences. Practise, practise, practise: the key to developing confidence when giving a presentation is to really know what you are going to say. Never read from slides with your back to the audience. Write all of the key points on small cards for reference, in case you lose your place. When you practice, time each presentation until you are confident that your timing is perfect. Seek the views of a friend or colleague. He or she may be prepared to attend a 'dress rehearsal' and provide an honest appraisal of the clarity and standard of the presentation.

If you are still feeling uncomfortable about making a presentation try the following: imagine that you have just finished your presentation, that people have been asking questions with genuine interest and that you have been confidently answering each one. Try to imagine how you will feel at this point, how happy you will be to have finished the presentation successfully and on a high note. Practise this technique and capture that very positive feeling as you go in to face your audience.

Your formal presentation skills are likely to improve markedly with each presentation. So, seize the chance to talk about your work and ideas whenever the opportunity arises; you will rapidly become a very effective communicator.

> *"It usually takes more that three weeks to prepare a good impromptu speech."*
>
> Mark Twain (1835–1910)

5.3.2 The elevator pitch

Pitching your idea(s) to someone in an informal setting can be every bit as nerve-racking, exhilarating and challenging as a formal presentation to an audience of peers or other interested parties. This section aims to prepare you for such a situation so that you are ready to communicate your ideas whenever the opportunity arises.

An elevator pitch is a brief overview of an idea that can be presented within approximately 1 minute. The term is usually used in the context of an entrepreneur pitching an idea to venture capitalists. It originated in the US where popular business culture recognises that meeting people by chance and for a very short time – say, in an elevator/lift – can present valuable opportunities for business. In the elevator pitch you have only the time it takes for the lift to reach its destination to sell your idea.

You can use the elevator pitch in any situation that involves a brief meeting, for example, when you meet someone influential at a conference, when you are networking with potential sponsors, or when you are simply 'testing-out' your ideas. Follow these rules and you will have a pitch ready to hand.

1. Think about your image – make sure you look 'business-like'.
2. Know your facts and figures, because not knowing the answer to a question about finance may sink your pitch. Make sure you are entirely truthful – if you are not you are likely to be found out.
3. Be aware of any weaknesses associated with your idea so that you are ready to deal with tricky questions.
4. Open with a statement that will act as a hook to catch the listener's attention.
5. Keep the message clear – don't use terms the listener may not understand.
6. Talk about the need your idea/product/service will meet and who will benefit – keep this short and sweet. Illustrate with statistics if appropriate.
7. Show your passion – if someone is thinking of investing in you they will want to see your enthusiasm and energy.
8. Close with a request – you may wish to ask the listener for another meeting, or whether you can send him or her further information, or simply whether he or she would like to support your idea.
9. Keep your pitch to 60 seconds, no more. Busy people have little time to talk to strangers.
10. Practise in front of a friend and ask for feedback. Practise, practise, practise.

An elevator pitch might go like this:

"My name is Susan Smith and my company, Watersafe, manufactures water testing kits, at a fraction of the price of any other kit on the market, using my patented technology. We are in the 21st Century and yet we have 1.1 billion people who have no access to clean water. I feel passionately about changing this situation but in order to be able to

manufacture large numbers of kits I need to invest in new equipment. This equipment will allow us to manufacture the kit at a price affordable in developing countries and the kit really will help people identify where it is safe to drink. So I am looking for an investor who is willing to invest £40 000 for a 10% share of the company. I appreciate your time in listening to me and I would like to be able to send you some more information on this. Do you have a business card that I could have?"

When you are speaking with others it is important to make sure that you say what you mean and understand fully what other people mean. Words have the power to evoke images, feelings and even a desire for action in the listener. Advertising copywriters rely heavily upon these effects, for example:

"*Just Do It*" - Nike

"*Guinness is good for you*"

"*It needn't be hell, with Nicotinel*" evokes an image that smokers recognise: giving up smoking is, apparently, hell

"*Beanz Meanz Heinz*" – creating new words that are still recognisable

"*You've been Tango'd*" – creating a verb from a brand noun

The important point is to develop an understanding of how you can use language to change people's perspective and persuade them that they would like to know more.

5.4 – Written communication

In the biosciences, our thoughts and ideas about a given subject, or area of investigation, are often conveyed by means of formal, written communications.

"I think writing skills are more important than presentation skills, at least for me. For me the written text is the most important and presentations are a means of improving your product, which is something written. For some people it might be the other way round."

Researcher, University of Wales

Written material can take many forms: reports, essays, memos, e-mails, journal articles, discussion boards, theses, research proposals, business plans, etc. All of these forms of writing have particular standards and involve specific ways of presenting thoughts and ideas. The format for a business plan is of particular importance to a budding entrepreneur and you will find detailed advice on how to prepare an effective business plan in *Chapter 7*.

5.4.1 Format for a scientific report

A widely used method of written communication in the biosciences is the scientific report and this section describes a structure and format for this form of communication. This should ensure your research results, product, service or ideas are clearly received and understood by your target audience.

Scientific reports can generally be divided into six parts: an abstract, an introduction, a methods section, a results section, a discussion, and a references section.

Abstract – the abstract should be a concise summary of the report. It should stand on its own, with no direct reference to the main text, and should indicate: the aim(s) of the work; the approach taken; the key findings; and major conclusions.

Introduction – the introduction sets the scene. It should indicate the purpose of the investigation and should contain a survey, and analysis, of the relevant background literature.

Methods – the key thing to remember about this section is that methods must be described clearly so that experiments or other procedures can easily be repeated by others. Details of equipment, materials and software used must be listed accurately. You should provide a detailed description of any new procedure developed during the course of the investigation. For established procedures, the reader should be referred to the relevant literature.

Results – the main text of this section should include a full description of the results obtained. The data may be summarised in figures or tables (typical examples are shown in *Figure 5.2* and *Table 5.1*) but avoid expressing the same data in tabular form *and* as a figure. Table and figure legends should enable full understanding of the data displayed without reference to the main text. Within this section, discussion of results should be kept to a bare minimum.

Discussion – in this section you must interpret your results using clear, logical reasoning. Your conclusions must be supported by the data you have described and related to the published literature. You should give an indication of the implications of the study and you may wish to make suggestions for future work in related areas.

References – you must provide a comprehensive and accurate list of scientific articles, reviews, textbooks and electronic sources of information (websites) consulted. The format you adopt will depend upon the nature of the document/publication you are preparing. There are many alternative reference styles for scientific publications. You must ensure that you conform to the format indicated by the publisher when you submit a manuscript for publication in a journal.

Table 5.1: Frequency of bacteriocins in a collection of *E. coli* faecal isolates from humans

Bacteriocin class	Type	Frequency (%) (n=266)
Colicins	Ia	9.0
	E1	8.3
	M	4.5
	E7	2.3
	K	1.9
	E2	1.1
	B	1.1
	Ib	0.3
	E6	0
	A	0
	D	0
Microcins	H47	13.2
	M	12.0
	V	5.3
	C7	1.5
	J25	1.5
	L	1.5
	B17	1.1

Source: Gordon and O'Brien[3]

Figure 5.2. VCH binding to sterols. ▶
VCH monomers (200 ng in 10 µl of TEN) were mixed with 500 µl of 3% non-fat dry milk dissolved in PBS and incubated with PVDF disks coated with 500 nmol of different sterols. Data from three independent studies that were performed in triplicate were averaged. Error bars show standard deviations. The VCH monomer did not bind to an uncoated PVDF membrane. Source: Ikigai *et al.*[4]

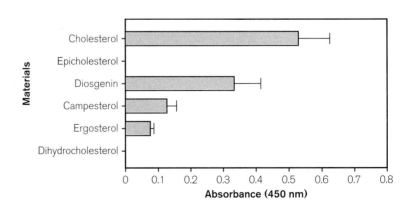

Scientific reports are written in the third person and enable the author to take on a more formal style than that employed in, say, a magazine article or newsletter. Writers preparing scientific reports therefore use language such as:

"Response surface methodology was applied to study the effects of fermentation on the levels of phytochemicals (folates, phenolic compounds, alkylresorcinols) and on the solubilization of pentosans in rye bran from native and peeled grains. Furthermore, the microbial composition of the brans before and after fermentation was studied[5]"

rather than:

"I used response surface methodology so that I could look at fermentation and solubilization in rye bran. I then found it easy to look at the bugs in the brans before and after my experiments."

Because scientific reports are formal and are usually detailed accounts of research or some other form of idea development, they tend to be valued more highly than many other means of communicating thoughts or ideas. However, a scientific report is not the *only* way to communicate your ideas in writing. Indeed, writing and submitting research or business proposals intended for audiences who are not scientists is likely to require a quite different approach. For example, if you are trying to convince regional or national funding agencies/policy-makers (such as Regional Development Agencies, local authorities, or banks) of the value of your idea, another style of writing will be more appropriate. Such research/idea development proposals can be varied in their scope and structure: one popular format is described below.

5.4.2   Format for a research proposal

Research proposals usually take the following form (adapted from Wilkinson[6]):

Title – the title of your work.

Introduction – this provides the context for your proposed work, with reference to the key literature sources and perhaps a refinement or clarification of the remit of your intended work.

Aims/objectives – the substantive elements of your work defining what you hope to achieve. Often, the aims/objectives section of a proposal is provided as a list of research questions, related to the title, which you wish to explore.

Methodology – the approach to the research that you intend to take, and the tools you intend to use within your research framework.

Analysis – an appreciation of the data you will collect, and indications as to how these data will be analysed. Here you might discuss particular computer packages or statistical techniques you will use with the data.

Timetable – the chronology of events/activities within your project work.

Dissemination – how and where you intend to share the results of your work with others, for example, through interim reports, workshops, conferences, journal articles, monographs, etc.

Budget – how much the work will cost.

Staffing and CVs – an indication of who will work on the project. This section should include a CV for the applicant and any named personnel with contact details and relevant previous experience such as: undergraduate and/or postgraduate project/dissertation details, research assistance work, seminar presentations, conference papers, other publications, etc.

References – the names (usually at least two) and contact details of those who can comment upon your academic ability and potential.

Scientific articles, and research and business proposals are often lengthy pieces of work. Many busy academics, venture capitalists and other reviewers will not have the time (or, for some, the interest) to read the entire document. Instead they will focus on key elements: the summary/abstract, the introduction, and the conclusions sections. Of these, the summary/abstract is especially important. It should be an accurate distillation of the entire document upon which it is based. You should take particular care with this part of an article, report or proposal because most readers will, initially, focus on it. If their interest is captured and maintained then it is much more likely that they will read the entire document.

5.4.3 Writing scientific reports and articles: some tips

Abbreviations and acronyms – wherever possible avoid abbreviations; when abbreviations or acronyms do have to be used, ensure that they are defined in full. You should avoid starting a sentence with an acronym or abbreviation. Some abbreviations and acronyms are used routinely and are generally accepted and widely understood by the bioscience community, for example, ATP, DNA, units of measure (m, g, °C), and mathematical or chemical formulae.

Contractions – when writing for academic purposes, you should not use contractions such as didn't, can't, haven't. Instead use the full words – did not, cannot, have not.

Data – the word 'data' is plural and is commonly mis-used. For example, 'the data *were* analysed ...' is correct.

Sentences – you should review your writing to make sure that each sentence presents no more than one or two ideas. Each sentence within a paragraph should sit comfortably with the preceding and following sentences. Ideas should be expressed clearly and in a logical order.

Scientific names – genus and species names should be underlined or italicized; only the genus name should start with a capital letter, e.g. *Homo sapiens, Ilex opaca.*

Significance – the word 'significance' is often mis-used. In academic writing 'significance' usually relates to the results of a statistical test. You should therefore avoid using the word (unless you are referring to data upon which an appropriate test has been performed).

5.5 – Posters

During recent decades, the poster display has become one of the most popular methods for communicating ideas. Posters are really useful for visually displaying information during scientific meetings. You should use a poster to convey your results and ideas clearly, concisely and with visual impact. A poster is very much like an elevator pitch: it must rapidly capture and retain the interest of the audience. *Figure 5.3* shows a poster designed for an enterprise learning conference and provided introductory information

Figure 5.3: Clear, uncluttered poster. ▸

about materials that had been developed for bioscience students at the University of Leeds. It has clear, uncluttered text with supporting imagery, and provides contact information should those viewing the poster wish to discuss its content further with the presenter/author.

During scientific meetings, poster presentations are often held in busy, congested areas such as corridors or foyers. There will probably be many posters in these confined spaces and it is likely that you will have to work hard to gain people's attention. You should also note that many of the conference or exhibition delegates will be attending the session to meet friends and network. Poster events are therefore usually accompanied by refreshments and nibbles. In other words, you will be competing with other factors when trying to display your work, so your poster needs to be good.

You are likely to spend only a small amount of time standing close to your poster. You must therefore design the poster so that it communicates your results and ideas without any need for further, detailed explanation from you or your collaborators. When you do stand by your poster you will have the chance to interact with people, to answer their questions, to ask them questions, and generally to network. If you craft your poster well, and talk effectively in support of your results and ideas, you will have the opportunity to impress a large number of people and make many useful contacts.

5.5.1 Creating an effective poster

Always make sure that you read carefully the information provided by the conference organisers and that you are aware of the precise dimensions of the poster. The following general guidelines are intended to help you create a poster that will really help you communicate your message. Your poster should be attractive, informative and memorable!

Consider the audience – it is important to match the contents of the poster with the needs of your audience. If you are trying to communicate a scientific issue to a non-scientific audience, then you will need to adjust your language accordingly. There is no need to patronize your audience by being overly simplistic, but you should explain scientific terms wherever you can. Avoid abbreviations except for those in everyday use, e.g. DNA. Think about the nature of your audience and convey your message in a manner that will effectively resonate with them.

Think carefully about the design – think about the size and shape of your poster and, if the organisers permit the option, whether it should be viewed landscape or portrait. Design a layout that will capture the attention of as many of the targeted audience as possible. It may help to prepare a draft layout using several sheets of A4 paper. You can sketch out your thoughts and ideas on these pieces of paper and move them around on a flipchart until you are happy with the design and can proceed to preparation of the final copy. Your poster should have a 'professional' finish. High quality posters can be created readily using PowerPoint software and your university should have facilities that will allow you to print out large, A1 poster-size documents directly.

Crisp and clear layout – the title/heading should sum up your idea and the essence of what you want to communicate. You should use a large, bold font and, normally, a maximum of around eight words that are easily read from a distance of two metres. The title must be snappy and informative – remember there will be lots of other posters vying for attention. Information about the authors, and their affiliation, should be indicated near to the title. For the remainder of the poster, the content will vary depending on the purpose of the communication. For a scientific communication, ensure your poster has a title, summary/abstract, introduction, materials and methods, results, conclusions, acknowledgments, and further information sections. Note, however, that if the poster is intended to communicate a different message then the layout and content may differ markedly from this formal format (for example, see *Figure 5.3*). Restrict the textual information to around 500 words. You do not need to fill all the available space on the poster; if you do it will look too 'busy'. All text should be double-spaced and should be easily read from a distance of at least one metre. By all means use colour, but use it sparingly. For example, you can use colour to link key aspects of the poster, or to link elements such as a graph with results described in the text. Avoid presenting 'raw' data – wherever possible, use simple graphs, charts and histograms to communicate your results. The golden rule is to keep the visual representations as simple as possible.

Conclusions section – people often read only the summary/abstract and the conclusions sections. You must make sure your conclusions are firm and clear. A list of bullet points within this section will ensure brevity and impact.

During the session your poster is there to sell you and your ideas. You should be enthusiastic and well-briefed. Respond positively to anyone showing interest and make sure you can answer any questions they may have.

5.6 – Communicating with non-scientists

Enterprising bioscientists must communicate effectively with members of the general public on a regular basis. As an entrepreneur you will wish to attract funding from individuals who lack a scientific background or education. Then, once you have established your company, you will need to communicate the nature of your bioscience-based product to prospective

Case study 5.1: The MMR controversy

Inaccurate reporting of research findings which may have serious repercussions for public health

The controversy that has surrounded the Measles, Mumps and Rubella (MMR) vaccine started in 1998, when Wakefield et al.[7] published a paper in *The Lancet*, which reported the cases of 12 children with developmental disorders. The authors proposed a link between a regressive autism condition and gastrointestinal signs and symptoms. Crucially, the authors also indicated that in most of the 12 cases, the onset of symptoms followed MMR immunisation. In fairness to the authors, they indicated in their publication that they did not prove an association between MMR vaccination and the syndrome described. However, during a press conference held prior to the publication of the paper, the leading author said he thought it prudent to use single vaccines instead of the MMR triple vaccine until the latter could be ruled out as an environmental trigger for autism. This generated a great deal of interest and heated debate in the scientific and popular press, and some of the newspaper articles mis-represented the findings of Wakefield and co-workers. Examples of headlines that contributed to public concern:

'Second MMR jab gave my son autism' – *The Sun* 18/10/2003

'Are we facing an autism epidemic?' – *Daily Mail* 13/03/2003

'Blair shouldn't be playing God with children's lives' – *The Sun* 11/02/2002

Intensive research studies involving large numbers of children, undertaken during the intervening years, have failed to identify a link between autism and MMR vaccination. As a result most medical experts recommend continued inoculation with the triple vaccine. Unfortunately, the controversy and apparently conflicting advice surrounding the MMR vaccine have confused many parents and very significant numbers have decided against vaccination of their children. Doctors are very concerned and warn of future epidemics of measles and other diseases because the protection normally afforded by 'herd immunity' may not be achieved.

customers. Furthermore, you will be expected to develop good relations with the public at large, ensuring that you communicate the ethos of your company by explaining clearly the underlying science and its safety. This section provides a checklist of important considerations to be taken into account if science is to be communicated effectively to a lay audience. Stories from the news media that sensationalised or mis-represented important scientific and/or medical issues leading to unnecessary public alarm are also included to highlight the dangers of getting it wrong (see *Case study 5.1*).

When communicating your ideas to a non-scientific audience use clear, unambiguous language and do not use provocative words such as 'always' or 'never', for example 'exposure to mosquitoes in Indonesia always results in malarial infection'.

A 1998 television interview involving a scientist from a Scottish research institute provides another good example of how not to report science in the news media. The scientist discussed the results of experiments which he claimed showed that genetically modified (GM) potatoes provoke a toxic reaction in the guts of rats. At that time the research had not been published in a peer-reviewed journal. Despite this, the results were reported essentially as proven facts in the news media. This lent weight to the argument, from some quarters, that GM products are bad for the health of humans and animals alike. Approximately one year after the original television interview, a peer-reviewed paper appeared in a reputable scientific journal. This article demonstrated clearly that GM potatoes are not toxic to rats. However, by then the damage had been done and many members of the general public were further convinced that the use of GM technology in agriculture is inappropriate and hazardous. For further details and discussion of the importance of peer review see Stewart[8].

However, some scientists are particularly good at communicating science to the general public. Dr Christopher Smith, from the University of Cambridge won the Biosciences Federation Science Communication Award for 2006. This is an award that recognises research-active bioscientists from UK universities and institutes who make an outstanding and consistent contribution to communicating science to the public. Dr Smith writes for the 'Naked Scientists', 'a media-savvy group of physicians and researchers from Cambridge University who use radio, live lectures, and the internet to strip science down to its bare essentials, and promote it to the general public' (www.thenakedscientists.com/).

Here is another excellent example of effective communication with the general public, in this case involving an explanation of a fairly complex scientific concept:

"When I was a child, the game of Chinese whispers was popular at parties. I expect you know it. The first player makes up a story and whispers it into the second player's ear. The second player whispers it in the third player's ear, and so on. By the time the message has gone around, the original tale of how Lucy's cat got shut in the coal shed might well be an account of a blue hat that got stuck on a goat's head. DNA plays its own version of Chinese whispers all the time. It is quite possible that there is no longer a single cell in your body that has exactly the same DNA sequence as that which was in you when you began your life as a fertilised egg."

Tom Kirkwood, Professor of Medicine, University of Newcastle
The Reith Lectures 2001, *The End of An Age,* BBC Radio 4

Effective communicators of bioscience are in great demand. Indeed, as you develop your communication skills you may be attracted to a potentially exciting and rewarding career in the news media.

5.6.1 The press release

One of the most effective ways to communicate your ideas to large numbers of people is to write a press release. Before you do this you should think carefully about the implications. Are you ready to release your idea to the media? Have you ensured that your intellectual property is protected? (see *Chapter 3*). Once your idea is in the public domain you will lose control over how it is used.

When writing a press release, think of the nature of the audience you would like to reach. If, for example, you would like your story to be published in a tabloid newspaper, you will need to tailor it accordingly – did you know that the average reading age of readers of the most popular newspaper in Britain is 9 years?

Often it is not easy to get your ideas or story into print; your press release will need to be well crafted. A busy journalist will want you to provide a press release that is complete in itself – it should not require further elaboration. So, stick to the facts and keep it simple. Avoid flowery language and jargon, and be as objective as possible; this can be very difficult if you feel passionately about your idea.

The timing of the release is also important. If you can tie it into a related and topical event, that may help. Alternatively, you could wait until the summer months, when news is notoriously 'slow', to publicise your idea.

As you construct your press release, make sure you can answer the following questions:

- *Who?* Who is the story about? Who does it affect? Who might be interested?
- *What?* What is the essence of your idea? What are you trying to say?
- *Why?* Why have you pursued this idea? Why is it a good idea?
- *Where?* Where will your idea work?
- *When?* When did your discovery happen? When will the results of your idea be available?
- *How?* How did this idea come about? How will it work? How will the public benefit from it?

The first ten words of the press release are the most important so make sure they are effective. Ensure that you provide as much contact information as possible: the individual who should be contacted, with their address, telephone, fax, e-mail and website details. In other words make it as easy as possible for journalists to understand and utilise your communication. They are then more likely to clearly communicate your ideas to the public.

The last word on communicating science to non-scientists should go to Melvyn Bragg:

"I'm fascinated by the fact that we live in a time when so many people are doing fantastic work, and thinking in areas which it's not remotely possible for me to keep up with and these people are prepared to talk about it. They're prepared to come on In Our Time and other programmes on Radio 4 and try and talk to the rest of us . . ."

Lord Melvyn Bragg of Wigton,
presenter of *In Our Time* and Chancellor of the University of Leeds

5.7 — Networking

"I think networking is something a lot of people don't like doing ... a lot of the time it can be superficial and it is hard work but I like the whole experience of that."

PhD student, University of Surrey

When you raise the subject of networking, many people think of aggressive sales people 'working' a room in order to identify the most likely prospects (or victims) for their sales patter. This perception of networking is becoming less and less prevalent in the world of business. Networking is about much more than merely 'selling' a product, service or idea.

Networking is primarily concerned with relationship building. As effective, well understood relationships develop, communication will be more direct, focused and efficient. In this section tools and techniques are described to enable you to become an effective networker who can communicate ideas in a professional and meaningful way.

The very thought of networking sends shivers down the spine of many people, but this need not be the case. A number of networking techniques, once learnt and practised a little, can help to reduce considerably or eliminate this fear.

The networking introduction

Effective networking begins with an introduction. Take a few minutes to think about how you will introduce yourself, your work and ideas to someone at an event. Your introduction should be succinct and, as a bioscientist, you will often need to explain what you do in lay person's terms. Ideally you should be able to say who you are and what you do in two or three sentences. This isn't easy – practise with some friends.

Of course, you will often have to break the ice at a meeting, conference or other event. Sometimes your peers or colleagues will introduce you to someone new, but often you will find yourself among strangers. There will be a lot of other people in the same position as you, so bite the bullet and say hello to the person sitting next to you. Begin your discussion with a fairly tame query such as: 'which department/section are you from?', 'have you been to other conferences like this one recently?', 'what do you expect to get out of today?', etc.

Show an interest

Good networkers show an interest in what other people have to say. The importance of this is often overlooked. You can show an interest by looking attentive and maintaining good eye contact with the person or people you are talking to. Maintaining eye contact during 50–75% of the time it takes to have a conversation is considered appropriate – much more or less than this might make the person you are speaking to feel a little uncomfortable. Constantly looking over the shoulder of the person you are conversing with, to see who else is in the room, is impolite, so try to avoid this. Engage in conversation, commenting on what the other person is saying, to rapidly build and develop rapport.

Give as well as take

A good networking exchange should be beneficial for both parties. Effective networkers are expert at providing, as well as absorbing, information and data. You should therefore ensure that you provide information that is useful to the person you are networking with. This can be achieved through listening carefully to what they are saying, and responding with appropriate and helpful additional information that you may have. You should also try to provide helpful responses to questions posed by your networking partners.

Remember

Proficient networkers have good memories. If you show that you remember things you were told in a previous network exchange you will prove not only that you were listening carefully but also that you valued what was said. Often, the things you will remember (and then recall at later meetings) will not be entirely business-related. For example, you may remember that a colleague was leaving for a holiday when you last met. This will provide a useful starting point for a conversation at your next meeting.

"Networking is an essential skill. Be friendly with everyone, particularly people in your own department. Go to events if you possibly can and build relationships with people – so if you ever need to call on these resources you can. You also need perseverance – keep at your work. Even if you are a bit anxious about whether it's going to work or not. You need to be very well motivated too – but this isn't really a skill – it just shows that you are interested in what you are doing."

Researcher, University of Leeds

Benefits of networking

There are many benefits associated with building a large network of contacts. Networking effectively with other scientists working in related areas, will enable you to become aware of current and recent developments in your specific field of bioscience. By sharing your thoughts and ideas with others you are also likely to receive free consultancy, with potentially useful advice about possible future directions for your work. In addition, it's often very valuable to hear about the experiences of others. When they tell you about the mistakes they have made, listen carefully – they may help you avoid the same pitfalls in your own work.

Good and effective networkers are always eager to meet new people and discuss work and ideas with them – a useful contact made today may open up opportunities to fund the development of your exciting research, product or service idea tomorrow.

Effective networking is particularly important in the early stages of a business. The next chapter describes the complex network of stakeholders that will develop as you start your business in the biosciences.

5.8 – References

1. **Luft J and Ingham H** (1955) The Johari window, a graphic model of interpersonal awareness. In: *Proceedings of the Western Training Laboratory in Group Development.* Los Angeles: UCLA.

2. **Dawkins R** (1976) *The Selfish Gene.* Oxford University Press, Oxford.

3. **Gordon DM and O'Brien CL** (2006) Bacteriocin diversity and the frequency of multiple bacteriocin production in *Escherichia coli. Microbiology,* **152**: 3239–3244.

4. **Ikigai H, Otsuru H, Yamamoto K and Shimamura T** (2006) Structural requirements of cholesterol for binding to *Vibrio cholerae* hemolysin. *Microbiology and Immunology,* **50**: 751–757.

5. **Katina K, et al.** (2007) Bran fermentation as a means to enhance technological properties and bioactivity of rye. *Food Microbiology,* **24**: 175–186.

6. **Wilkinson, D.** (2005) *The Essential Guide to Postgraduate Study.* Sage, London.

7. **Wakefield AJ et al.** (1998) Ileal-lymphoid-nodular hyperplasia, non-specific colitis, and pervasive developmental disorder in children. *The Lancet,* **351**: 637–641.

8. **Stewart CN** (2003) Press before paper – when media and science collide. *Nature Biotechnology,* **21**: 353–354.

5.9 – Additional resources

Barrass, R. (1995) *Scientists must write: a guide to better writing for scientists, engineers and students.* Chapman & Hall, London.

http://classweb.gmu.edu/biologyresources/writingguide/PracticalTips.htm
Provides useful guidance on writing scientific reports and articles.

Learning outcomes

Key learning points from this chapter are:

- that the way we communicate our message, using visuals and language appropriate for the audience, is crucial for the effective development of our thoughts and ideas

- how to feel comfortable about presenting to a variety of audiences; and that much of this relies on practice and familiarity with your material and presentation tools and techniques

- that ideas that are readily communicated are those that can be effectively 'pitched' to others in a short space of time

- how to write effective scientific reports and research proposals that are informative to the reader and demonstrate your knowledge of the subject matter

- how to create an effective poster using design, an appropriate format and structure, and a careful consideration of the target audience

- how to communicate effectively with non-scientists, in language and terminology they are familiar with and understand

- that networking is a fundamental business (and life) skill that requires practice, practice and more practice

Chapter **6**
Starting up a Business

Alison Price & Ted Sarmiento

6.1 – Introduction

When you decide that you want to start up a business, you will find yourself confronted by a wide range of practical issues including insurance, premises, and health and safety. Many of these issues require up-to-date advice as they vary depending upon the nature of the business and other key factors, including where you choose to base the operation. However, by far the most important consideration for any business start-up is the nature of the people involved. Investors often readily admit that they invest in people not ideas because, crucially, business team members bring the product to market, overcome difficulties and support one another to develop a successful business. Equally, a team with limited experience, or lacking specific business skills, can effectively limit the potential of a solid business idea or new technology. We therefore focus here not only on some of the legal issues and cost considerations of starting up a business, but also upon the roles of people within the organization. This should help you gain an overview of the whole process of starting a new business.

This chapter builds upon the skills explored and developed during preceding chapters and aims to instil an understanding of business start-up in the biosciences. The range of options available to potential entrepreneurs starting a new business will be explored. Some of the key decisions that need to be made before drafting a business plan or starting a company will be considered in a way that should help you appreciate the practical and infrastructure issues that are required to create a viable business proposition. This chapter therefore lays the foundations for the next chapter that covers the business plan. However, note that regulatory frameworks (such as health and safety laws; see *Chapter 9*) and societal expectations shift over time, and so new entrepreneurs should not rely solely on the material presented here. You should always seek out the latest relevant information available from private, government and other appropriate sources.

6.2 – Building the team

There is no shortage of new businesses based on novel products or ideas but, as the Biotechnology and Biological Sciences Research Council (BBSRC) suggest: "creating the company is the easy bit. Turning it into a viable business is the tricky part!" (www.bbsrc.ac.uk/biobusiness_guide/). When starting and building a successful business it is essential to understand the nature of the team needed to 'make it happen'. This might mean expanding the initial development team (those who originally had the idea) to incorporate new members with additional business skills. This can be a big step for those anxious about sharing a new idea, but it can also make the difference between success and failure.

From the outset it is essential that roles are clearly defined for all of the team members who will contribute to the development of the new business. This is particularly important during the start-up and initial growth phases of a small- or medium-sized enterprise (SME – defined by the European Commission as a business with 10–250 employees and a turnover of 10–50 million Euros per annum; ec.europa.eu/enterprise/enterprise_ policy/sme_definition/index_en.htm) when members of staff are likely to be asked to adopt more than one significant role. *Table 6.1* summarises the key roles within a developing SME. The list is not exhaustive but helps clarify the range of responsibilities that should be considered at start-up.

Table 6.1: Roles within an SME (adapted from Lang[1])

Title	Role
Chair	Chairing meetings and checking minutes; brings experience and contacts; used for major dispute resolution
Managing Director (CEO)	Responsible for overall development and running of the enterprise, functioning of the team, and delivery of results. Formally responsible for the day-to-day running of the company, formulating policy proposals and implementing the board's decisions. Usually has a marketing rather than a technical background and is often not a founder member
Finance Director (CFO)	Responsible for financial calculations and analysis for strategic development. Keeps the statutory books and is usually a qualified accountant; prepares management reports and budgets; advises on fundraising; may also act as company secretary (keeping board minutes, official papers and shareholder records)
Technical Manager	Product development, refinement and improvement
Production Manager	Running the factory and the distribution chain
Marketing Manager	Deciding what and how to sell. Market communications (public relations, advertising, website, etc). Market and competitor information
Sales Manager	Selling; customer relationship management, including after-sales support
Human Resources Manager	Deals with people issues such as recruitment, promotion, remuneration
Company Secretary	Ensures the governance of the organisation, i.e. effective record keeping, meetings, minutes, and information flow

This is rather a long list of formal roles for most start-up businesses. Indeed, many businesses start up with only one person adopting essentially all of these roles! However, if you are to give proper consideration to the requirements of your start-up business, then you must think about all of these roles so that no area is over-looked and the route to business expansion, with the appointment of new staff, is identified. This is particularly important for bio-businesses which are often started by a number of individuals rather than just one person. Successful enterprises of this nature are dependent upon a clear understanding of roles within the team.

In addition to the roles indicated in *Table 6.1*, Lang suggests that a scientific advisory board should be appointed to enhance the credibility of a new bio-business[1]. While the scientific board will not have a legal role within the organisation, it will be able to advise on the scientific direction of the company and will be a useful method of engaging senior figures from a particular area of science in the start-up venture. Friedman[2] also emphasizes the value of a scientific advisory board claiming that "young biotechnology companies have three basic team components – scientists conduct research, management provides direction and administrative support, and a scientific advisory board provides an objective assessment of the company's progress and goals".

Such external support can mentor and guide a new business but, as academic and entrepreneur Alan Kingsman advises: "having a team that you trust and where you are honest and frank with each other is absolutely essential" (www.bbsrc.ac.uk/biobusiness_guide/). This sentiment is echoed by O'Donovan:

"Find a partner or partners whom you trust and get along with (you will be spending much time with them over the next few years) to share the thinking, planning and other responsibilities associated with starting your company. Two people make a sufficient starting team; three is good and more than four is probably too many. Creating a company and attracting funding to it can be a lonely task, especially for the first time. Core skills required in the first year will be creating the 'product', followed by identifying customers and making sales. These skills must be clearly visible in the founding team. Leadership is crucial and if an obvious single leader does not appear within the founding team, then he or she must be recruited before or shortly after first-stage funding is secured. It is also important to identify the remaining key management gaps and agree to fill them within a reasonable period. The skills required to grow the company change with the growth trajectory of that company and may be added as needed"

(www.enterprise.cam.ac.uk/building/starting.html).

However, for every piece of advice there is often someone else whose experience advises differently. While in the example above O'Donovan advocates limiting the number of individuals in the core team to no more than four, Southern and West[3] argue that to help ensure success, all start-up companies should, from the outset, adopt a model of a team of five key individuals who are at the very heart of the enterprise. In addition to the *entrepreneur* who is the driving force, the person with both vision and charisma (the 'boss'), Southern and West suggest that four other important roles should be adopted by individuals of suitable calibre:

- the **technical innovator** – the person whose invention or innovation is at the heart of the enterprise
- the **delivery specialist** – someone who can get the product/service to the customer, someone who can 'deliver'
- the **sales specialist** – someone who will find customers, if there are customers to be found;
- the **financier** – someone who understands corporate finance inside out, and will keep cost under strict control

These five key people map easily against the roles outlined in *Table 6.1* and provide a decision-making team which can, as team-guru Belbin suggests: "balance well with one another". He argues: "the useful people to have in a team are those who possess strengths or characteristics that serve a need without duplicating those that are already there"[4].

In summary, it is clear that a successful start-up requires detailed consideration of the roles within the organisation. It is essential to ensure that the personal strengths, skills and abilities of individuals effectively complement those displayed by other members of the team.

6.3 – Stakeholders

Having recognised the key importance of the team, there are others who should be taken into account when thinking about starting a business. Apart from the mentoring and advisory role of a scientific board, there are other, important groups of stakeholders who must also be considered.

Stakeholders are individuals or groups who, quite literally, have some form of stake or interest in the new business. They are the individuals or groups who will start, staff, run, support, develop, promote, sell to and buy from the growing company. When a business is being established, it is important to consider the full range and nature of stakeholders in some detail in order that interactions with these groups of people, and their contribution to the growing organisation, may be managed most effectively. The following should be considered:

- shareholders/owners/investors
- customers
- employees
- suppliers
- societal interests
- universities/academic institutions
- regional/local development agencies

Many of these stakeholders will demand information and expect high standards of business behaviour. Failure to acknowledge the needs of these groups can result in hidden costs and damage to the business. Whilst it is said that there is no such thing as a bad press, not managing publicity can result in damage to the company, its reputation and sales. The following sections explore in detail if or how these stakeholder groups will be important to your new business.

6.3.1 Shareholders/owners/investors

The exact make-up of this group will largely depend upon decisions relating to organisational structure and investment funding discussed in *Chapter 8.* Whilst all investors are keen to see a return on their investment (so that the money they have put into your company earns them more than if they had put it in a bank) investors can have other concerns, such as social and ethical issues. So some investors will want to align their investment with their ethical views – just as customers who bank with the Co-operative Bank (see *Case study 6.1*) want to align their banking (investing or daily transactions) with that organisation's ethical code (for more detailed consideration of ethical issues, see *Chapter 10*).

Case study 6.1: The Co-operative Bank
"Customer led, ethically guided"

The Co-operative Bank, which was founded under co-operative principles in 1844, launched an ethical policy in May 1992, under the following philosophy "At the Co-operative Bank, we always remember that it's your money in your account. Our role is simply to take good care of it for you – and not do things with it that you wouldn't do yourself."

This resulted from a major ethics survey of 30000 customers in which the majority (84%) of customers who responded believed that it was a good idea for the Co-operative Bank to have a clear ethical policy. Their new policy statement was endorsed by 78% of customers and a further survey in 2001 has shown that this support had increased to 97% of customers.

It is also important to appreciate that different groups of investors will have differing expectations, so some shareholders or investors will be seeking a quicker return on their money than others. For example, it is unlikely that an investor who judges performance over the short term (less than 5 years) will invest in a research-based bio-business or spin-out, as several years of further research and testing may be required before the product goes to market and any return is seen. Such a business will need a different, more patient type of investor.

The majority of businesses that spin out of university research are unlikely to make any return in the first 5 years and this must be considered in the development of any business or investment plan. University spin-outs in the UK can be the beneficiaries of some additional help, as the UK now follows the US model of establishing technology transfer or business development offices within the university campus to assist with commercialisation of new ideas. Nelson[5] suggests that in the US, experienced technology transfer offices and a vibrant business community, experienced in specialised start-ups, can "provide additional resources to support the birth and growth of new companies", where "there is more venture and angel capital willing to invest in early-stage companies and early-stage technology and that angel and venture investors are more competent to guide the formation and early growth of companies". Over the last 10 years, we have seen many UK universities start to provide this sort of expertise in support of the development of businesses emerging from bioscience and other research departments (see *Chapter 8*).

6.3.2 Customers

Customers expect businesses to provide them with goods and services. Within the trading relationship that is established between individual customers and the organisation, there are often many implicit or unspecified conditions. These are implied assumptions about quality, durability, performance, safety and other factors which need to be understood by both parties to ensure a successful transaction. In this regard, it is important to be aware of the service and quality offered by your competitors so that you know what your new customers will expect and ensure that you meet, or even exceed, their expectations (the concept of 'competitive intelligence' is discussed in detail in *Chapter 4*).

Most bio-business customers are likely to be other organisations (universities, the NHS, etc.) or large companies. Whilst this should provide clarity of expectation through the drafting of legal contracts with clear product specifications, often the additional people involved within the businesses can add further demands. Business-to-business (B-2-B) sales are conducted by organisational buyers, which can mean that there is a chain of individuals involved in each purchase, rather than one person. This chain can include a range of people, with different roles, which might mean that the product is approved by technical experts for functionality, but the sale is finally sanctioned by finance, who might first require legal and financial assurances. Most importantly for a new business, it must be appreciated that sales from one business to another may involve significant delays while invoices are raised, processed and finally paid. Many businesses do not settle bills in less than three months and some large organisations can take longer to send payment. Such delays can have a significant impact upon the cashflow of a small business and many SMEs go bankrupt simply because of difficulties with cashflow. These cashflow difficulties affect all team members, and employees will soon recognise the need to ensure that their regular monthly direct-debits for mortgage, bills and other personal payments are not planned for the first day of the month as the cash may not be available at the start of the month to enable the new bio-business to pay salaries.

6.3.3 Employees

As well as gaining employment from an organisation, employees also have a range of implicit work needs such as job security, safe working conditions, engagement in rewarding work, fair treatment and recompense, and personal and professional development. The management of staff places legal as well as ethical costs and concerns on employers. These include: equal opportunities, promotion practices, employment continuity, redundancies, remuneration, trade unions, working conditions, skill development and training, and even drug/alcohol abuse support. Some of the costs associated with these issues will not be easy to plan for. However, some critical development costs, such as training for new skills, advertising and recruitment of new staff through expansion can be built into a business plan (see *Chapter 7*).

6.3.4 Suppliers

Suppliers provide the raw materials or components for new products or services. They also work with a new business at the B-2-B level and, as with any other organisation, have expectations of the businesses with whom they trade. These expectations are often legally enshrined within a contract and can include the due date by which the supplier expects to be paid (they will often expect you to pay earlier than you can obtain payments due to you from larger enterprises, thereby exacerbating your cashflow problems). It is important to agree all of these expectations in writing, usually within a contract, to avoid confusion, disappointments or delay. Additionally, suppliers often seek further clarification and confirmation from fledgling businesses that they will be able to pay. Ideally, business success is accompanied by the development of trust and the other benefits of a long-term trading relationship between businesses. However, this can only be achieved once credibility of operation (reliability, sustainability, etc.) has been established over time.

6.3.5 Societal interests

Many issues and agendas engage consumers, as members of society, and influence their purchasing decisions. Closely related to this is the role that the news media play in identifying and building awareness of new issues such as animal or human rights, global warming and carbon 'footprints'. It has been increasingly recognised that the quality of all of our lives will depend on how well businesses juggle the often conflicting notions of 'profit' and the 'environment'. As a result, the environment has, over recent years, become an increasingly important issue for business. This has been prompted by consumer interest in the environmental impact of corporate operations and even those selling direct to other businesses now find they must respond to these concerns. A new business venture may be able to incorporate these concerns as a positive outcome of its proposed activities (see *Chapter 10*). For the bio-business, additional societal issues may arise. Thus, a new bio-business must take into account how society perceives its activities in relation to issues such as animal testing, nanotechnology, and the general approach of the news media to reporting new developments in science and technology.

6.3.6 Stakeholder management

From the list of stakeholder groups above, you will now appreciate the wider world in which your business will operate. An awareness of stakeholders is essential from the earliest phase of planning a business. You need to consider the impact these groups may have and how their interests might support, or conflict with, those of your business. You should start by creating a list of stakeholders for your own business, beginning locally and building up to those stakeholders with wider social and global concerns. You then need to think about how you can work effectively with these stakeholders. It can be useful to divide stakeholders into groups according to the views or positions they hold in relation to a specific problem or issue. You might, for example, divide them into groups that are 'internal' or 'external' to the company, or into 'primary' (directly involved or related to the activities of the business) or 'secondary' groupings, to help you identify the group whose views should carry most weight. Creating this list and assessing the importance and impact of these stakeholder groups are important first steps as you devise strategies for actively responding to each group.

One technique to help you develop these strategies is called 'stakeholder mapping'[6]. This allows you to think through the long-term activities and demands of all of your stakeholders and recognise how these might change over time. It will alert you to potential problems and help as you develop suitable strategies to address emerging issues. The approach is particularly important for companies in the biotechnology sector who must frequently deal with highly emotive and controversial issues. A matrix is created (*Figure 6.1*) and each group of stakeholders is analysed to determine their relative levels of power, influence and dynamism in relation to the activities of the new business. This helps manage what may be a very wide range of stakeholders, assists with public relations and marketing, and ensures that the new business is established effectively and can operate over the longer term. Three specific judgements need to be made:

1. how likely it is that each stakeholder group will impress its expectations upon the company
2. whether they have the means to do so
3. the likely impact of these expectations

The stakeholders may now be placed into one of four boxes in the matrix with regard both to their power and predictability. Clearly, highly unpredictable and powerful stakeholders will need more monitoring and management than those who are predictable and have less power.

Figure 6.1: Stakeholder mapping – the power/dynamism matrix ▸
Adapted from Johnson and Scholes[6].

		Predictability	
		High	**Low**
Power	**Low**	Few problems	Unpredictable but manageable
	High	Powerful but predictable	Greatest danger or opportunity

This matrix will allow you to rapidly identify the most powerful and potentially troublesome stakeholders and, perhaps, anticipate and deal with problems. For example, the results may prompt you to create a public relations campaign to help ensure general acceptance of what might otherwise be a controversial new product or service. The analysis will require regular review and reconsideration, both with regard to the list of stakeholders and the categories in which they have been placed.

Public attitudes to the emotive issue of animal testing illustrate how a situation can change with time. Any business working with animals, or with materials tested using animals, needs to monitor and consider the views of the general public, and other stakeholders, and respond accordingly.

The use of animals for testing cosmetics and other products was raised as a consumer issue by *The Body Shop* more than 25 years ago. They used the fact that their products were not so tested as a very positive and effective marketing strategy. Consumer pressure built and resulted in changed practices by many companies, including those outside the beauty sector, as shoppers demanded that their products should not have been tested in procedures involving animals.

Stakeholder mapping by the *The Body Shop* at around 1980 might be represented by the matrix shown in *Figure 6.2*. At that time the views of shoppers, once aware of the issue of animal testing, might have been perceived as having low power and high predictability. News media were less predictable as were potentially powerful and radical animal rights activists. For a considerable period, updated matrices constructed by *The Body Shop* would have closely resembled the matrix shown in *Figure 6.2*. As a result, *The Body Shop* adopted a similar sales and communication strategy for many years.

Figure 6.2: Stakeholder mapping for animal testing c. 1980 ▶

Predictability

		High	Low
Power	**Low**	Members of the general public	News media
	High	Governmental organisations	Radical animal rights activists

However, in recent years attitudes appear to have shifted. This may be due, in part, to the extreme and occasionally violent activities of some animal rights activists. The shift in public opinion was perhaps most apparent in a recently published statement, signed by a host of luminaries including Prime Minister Tony Blair, in favour of certain forms of animal experimentation. As a result, new groupings have been created and must be managed (*Figure 6.3*). A company contemplating animal testing at the present time must take into account the contributions of influential

Figure 6.3: Stakeholder mapping of animal testing in 2007 ▶

Predictability

		High	Low
Power	**Low**	Members of the general public	Corporate PR campaigns
	High	Governmental organisations	Radical animal rights activists; individual high profile 'celebrities; media response

celebrities who may be highly vocal in their opposition to, or support of, animal testing. The response of the news media to these individuals, and other issues, can be both powerful and unpredictable. A new business may therefore wish to consider the implementation of a PR campaign in support of its activities.

Repeated use of the stakeholder mapping approach will allow a new start-up company to consider both how groups are changing and the response they should make to these changes. At first sight it may appear that a new, emerging business has little to contribute to wider social and national issues. However, as many consumers 'think global and act local' it is important for any business to appreciate the issues that their stakeholders care about. For example, as the head of a start-up company you may wish to decide on a response to key international issues, such as 'the size of the company carbon footprint' and decide whether this is likely to be an issue with your stakeholders. If so, you need to determine appropriate strategies for stakeholders in each of the matrix categories to manage your business effectively for the longer term.

6.4 – Options for bioscience enterprise

To avoid potential loss of IP, many universities are now encouraging research staff to consider opportunities for IP exploitation (see *Chapters 1 and 8*) and this may mean that researchers are asked to delay or avoid publication of some of their work. The main reason for this is that public disclosure of new ideas can mean that it is then not possible to go on to patent these ideas (see *Chapter 3* for full details of patent protection).

Ideas, discoveries and innovations can be exploited or commercialised in a number of ways. Whilst starting a business is the obvious choice for many new entrepreneurs, it may not always be the best approach when commercialising an idea within the bioscience industry. Therefore, before looking at the different legal forms that a new start-up business can take, it is important to consider other options that are available, including: licensing; selling; the establishment of strategic alliances; or the creation of a spin-out (as a form of start-up) company. *Table 6.2* provides a summary of these options.

6.4.1   Licensing

A licence is essentially a legal form of 'permission' for a third party to utilise IP for a specific purpose. This could mean that all financial risks are assumed by the third party with the inventor or creator receiving income from the successful commercialisation of the IP. Licences can take a variety of forms:

Table 6.2: Overview of enterprise options

Commercialisation strategy	Description	Advantages	Disadvantages
Licensing	Business model is to gain income by licensing use of technology	Spreads risk	Lack of control
Selling	IP is sold on to others	Little risk, immediate reward	Loss of control over application, potential loss of higher reward
Alliances	Strategic agreement with other company to build market share and develop business	Spreads risk	Legal constraints and joint decision making
Spin-out	Research is 'spun-out' of the university into a company format	Research is applied; benefits customers and profit can be made for research staff	Possible to lose focus and control of research in favour of new company

- they may be industry-specific, meaning that the IP can be used only in a particular industry; for example in the pharmaceutical, but not the automotive, industry
- they could be limited by geographical market (in a similar manner to franchising) meaning, for example, that the licensee may buy the right to use the IP for sales in the UK and Europe, but not for sales in the USA or Asia

There are many variables to consider when opting for the licensing route to market. The main one will be the type of license to be granted which can be 'non-exclusive', 'exclusive' or 'sole'. An exclusive licence means that the licensee is the only party permitted to use the IP, whereas non-exclusive indicates that other licensees may also be given permission to utilise the idea or technology. A sole licence permits only one licensee to exploit the IP but the licensor also retains the right to use the idea/technology. Other key issues that will need careful consideration are:

- who exactly the licensor(s) is(are)
- the timescale of the licence, when does it start, when does it finish, and under what circumstances may the license be terminated (by either party)?
- what exactly is being licensed?
- what can be done with the technology and are there any restrictions on the use?

- where can the technology be utilised, i.e. any geographical, market sector or other constraints on its use?
- the payment terms: what payments will be made, when, and on what basis will they be calculated?

Hughes[7] argues in favour of licensing instead of setting up a company as follows: "in the five years it takes an entrepreneur to develop a 'wet' bioscience idea into reality, the opportunity has gone since science moves on or the idea has been coincidentally brought to market in six months by a company which put 200 staff to work on its development". Clearly, it is advisable to seek professional assistance when dealing with such matters in order that potential pitfalls of this nature may be avoided.

For further discussion of licence agreements, see *Chapters 1* and *3*.

6.4.2   Selling the technology

This may be the most appropriate option in some cases. For example, where the inventor has no continuing interest in developing the technology, or where the institution or inventor has neither the resources nor the inclination to fund or maintain patent protection, or where the prospect of an immediate cash receipt is more attractive than a longer-term, albeit perhaps

Professor Chris Lowe
Head of Cambridge University's Institute of Biotechnology (IOB)

"Taking something through from concept to commercial product is a long and tortuous path, whichever exploitation route you take. That's particularly so if, like IOB, your focus is on long-term research. So you need to be very determined and thick-skinned. Rejection is par for the course, but there are ways to increase your chances of securing an industrial partner. It really boils down to doing your homework and getting the people issues right.

Our funding comes from government, charities, the research councils and, of course, industry. Given our research profile, we look for companies who can take a long-term perspective, and who understand how our work can fit in to their long-term objectives. Some companies are so focused on their immediate needs that they really only want universities to do near-term research for them on the cheap. I'm afraid that many UK companies fall into this category.

You need to identify an individual within the target company you feel you can work with. Good personal relationships are, I think, the bed-rock of every successful research collaboration. That person has to be at the right level in the organisation. In general they need to be reasonably senior, but not necessarily at the top. If they are they probably won't appreciate the science. If he or she is too junior, they may understand the science, but probably won't have sufficient appreciation of the commercial and political issues nor the financial clout to push the deal through. You need someone who has a good understanding of both your science and the direction the company is taking to champion your cause inside the company."

Source: www.bbsrc.ac.uk/biobusiness_guide

higher, return. This option can be relatively straightforward and relieves the researcher of any ongoing involvement with the technology. However, there may be no control over what then happens to the technology once it has been sold.

6.4.3 Strategic alliance

If a new technology requires external expertise for further development then its commercialisation may be best achieved through the formation of a strategic alliance with an industrial partner (see *Case study 6.2*). Such an alliance may mean that the industrial partner is willing to assume much of the financial risk in return for a greater share of the future rewards. The original creator may be able to benefit from lower levels of involvement, but may also give up some degree of control over the commercialisation of the technology. The terms of an alliance, such as the ownership of the IP and revenues on commercial outputs, should all be carefully considered before any agreements are signed.

Case study 6.2: Oxford Diversity
Example of a strategic alliance

In 1992, Stephen Davis, then an Oxford Professor, secured £500 000 investment from two business angels to pursue a research project involving the small-scale manufacture of certain chemicals. He set up an SME with four employees to provide a one-stop shop in chemical discovery, as an outsource partner to drug manufacturers. In 1995 a joint venture was formed with the pharmaceutical giant Pfizer (who developed Viagra) to create Oxford Diversity, and in 1998 the company was floated on the stock exchange with a staff of 200 (40% PhDs).

Adapted from Burns[8].

6.4.4 Spin-outs/Start-ups

University employees are often asked to consider starting a spin-out business to raise income for the inventor and the institution (see *Case study 6.3*). These spin-outs can take a variety of forms, with differing levels of involvement for the key researchers. The spin-out strategy can allow university staff to work almost full time on the new business or simply to provide technical advice to the company; in each case members of staff retain university employee status (see *Case study 6.4*). Currently, many universities have supportive and generous packages that enable the creation of spin-outs as part of their knowledge transfer programmes (*Chapter 1*).

Case study 6.3: Orla Protein Technologies
A successful spin-out

"As the biosciences are becoming more quantitative and predictive, and more and more reliant upon insights from the physical and mathematical sciences to provide the analytical power to make sense of all new data, so new opportunities for innovation are appearing at the interface between scientific disciplines and in new areas such as nanotechnology" Doug Yarrow, BBSRC Director Corporate Science in 2004.

Jeremy Lakey and Dale Athey of the University of Newcastle recognised the potential for a spin-out from Lakey's research on dynamic aspects of life at the molecular level. They sought to commercialise a technology that exploits self-assembly procedures to produce functional protein-based surfaces. The approach has applications in areas ranging from drug discovery to medical devices and diagnostics. Lakey and Athey's proposal for Orla Protein Technologies was runner-up in the 2001–2 BBSRC Business Plan Competition, and the company has since gone from strength to strength.

Source: www.bbsrc.ac.uk/business/biu/
biosci_innov.pdf

Case study 6.4: Alan Kingsman
Academic turned CEO

Changing roles
"If you are going to leave your academic position to work full-time with the new company you have to be absolutely single-minded and determined to do it! It's very exciting but it's extremely challenging. It can be infinitely more stressful than working in an academic environment. You cannot have any doubt in your mind that this is what you want to do. That is the single most important consideration."

Motivation
"We were actively engaged in transferring our technology into the commercial arena. Gradually we became somewhat frustrated with what other people did with our technology once it was passed to them!"

Source: www.bbsrc.ac.uk/biobusiness_guide

6.5 – Legal forms of business for a start-up company

A spin-out, or any other type of start-up company, may adopt one of several legal forms. It is essential to decide, from the outset, which is the most appropriate legal form for the business because, while it is possible to change the format of the start-up after trading has started, any change will take time and cost money. As indicated earlier, the bio-entrepreneur should also consider the possibility of commercialising, without actually setting-up a new business, through licensing or the creation of a strategic alliance. Here the costs and benefits of starting up are explored, and some of the key decisions considered.

The new business may structure itself under one of several formats:

- sole trader
- partnership
- limited company
- limited liability partnership
- workers' co-operative, community interest company
- franchise

and each of these are considered in more detail in the following sections.

6.5.1   Sole trader

As a sole trader, the owner is self-employed, and there is no specific legal structure. The advantages of being a sole trader include independence, ease of set-up and running the business, and the fact that all the profits go to the owner/manager. The disadvantages include a lack of support and unlimited liability (the owner/manager is personally responsible for any debts run up by the business).

6.5.2   Partnership

In a partnership, two or more self-employed people work together as partners and share the profits (or losses). As with a sole trader, the advantages of being in a partnership include the ease with which it can be set up and maintained. In addition, partners can bring a variety of skills and expertise to the business. On the other hand, once again disadvantages include unlimited liability, and the fact that problems can occur when there are disagreements between partners.

6.5.3   Limited company

A limited company is a separate legal entity, distinct from its shareholders, directors and employees who therefore have limited liability. Unlike a sole trader or partnership, the legal entity is not the same as the individuals who own or run it. For example, the legal entity can sue or be sued in its own name. This is the most obvious form of business that will be used within biotechnology. Parks[9] suggests the two main reasons for choosing this form of business are protection and professionalism. Protection of all those associated with the company is assured, at least to a degree, as it is the separate legal identity of the company that will deal with debts, litigation and other legal obligations. A limited company presents a professionalism that leads to confidence about the business that is important to customers and suppliers, who may themselves be limited companies, as they will expect to deal with another such company. The 'limited' status may also make the new organisation seem more established and successful to potential customers, investors or other stakeholders.

A limited company may be private or public. A private limited company may not offer shares to the general public and is required to have the suffix 'Limited' ('Ltd') as part of its name. A public limited company (PLC) can publicly trade shares on a stock exchange. Limited companies must be registered with Companies House and comply with the requirements of the Companies Act 1985. This means that directors have personal responsibility to ensure that statutory documents are delivered. These include: accounts; annual returns; notice of change of directors or secretaries, or change in their particulars, for example, notice of change of registered office.

6.5.4 Limited liability partnership

A limited liability partnership (LLP) has some of the advantages (and disadvantages) of both a company and a partnership. The key features of an LLP are that it: has a separate legal entity distinct from its members; can own and hold property, employ people and enter into contractual obligations; has members but no directors or shareholders; has no share capital; and is not subject to the company law rules governing the maintenance of capital. The members of an LLP have limited liability but the LLP is liable for all its debts to the full extent of its assets. The organisation has flexibility as to the internal structure it wishes to adopt, but it must maintain accounting records, prepare and deliver audited annual accounts to the registrar of companies, and submit an annual return in a manner similar to the procedure adopted by a limited company.

6.5.5 Workers' co-operative; Community Interest Company

A workers' co-operative is not an obvious choice for a commercial bio-business as there is no typical start-up team; rather, the business belongs to the employees and operates through joint decision making. In July 2005 a new trading form – the Community Interest Company (CIC) – was introduced. CICs give social enterprises the flexibility of the limited company format, but with additional features to ensure that they operate for the good of the community, not simply for private gain.

6.5.6 Franchise

Finally, it should be noted that many businesses, particularly some fast food and other high street outlets, are *franchises*. Franchising can be described most accurately as a way of doing business, rather than a legal structure, as it is the initial or original business model, developed previously, that is being franchised or 'copied'.

Legal options for start-up companies are summarized in *Table 6.3*.

Table 6.3: Summary of legal options for start-ups

	Description	Advantages	Disadvantages
Sole trader	Sole owner who can employ others	Easiest to establish	Owner personally responsible for business losses (without limit); no income if on holiday or sick
Partnership	Joint ownership across partners	Spreads risk; more attractive to investors; broader base of skill, capability and knowledge	All partners responsible for actions of others; needs legal considerations (has cost implications)
Limited company (Ltd)	Creates separate legal entity; owned by share-holders; company has a legal identity	Clear structure for seeking funding; most attractive to investors	Higher set-up costs; accounts are public (lodged with Companies House)
Limited Liability Partnership (LLP)	A half-way house between partnership and limited company, mainly used for firms of professionals such as solicitors	Separate legal entity; can own and hold property, employ people and enter into contractual obligations	Liable for all its debts to the full extent of its assets
Co-operative, Community Interest Company (CIC)	Joint, equal engagement by all	Shared ownership and co-operation	Slow decision making; spreads financial reward
Franchise	Branches of same business with different ownership of each	Comparatively fast growth; spreads risk	Investment required to establish model

6.6 – Phases of business growth

In the UK, only approximately 70% of businesses can be expected to survive beyond the first three years (www.sbs.gov.uk/survival). It is therefore important to try to understand the phases of growth for a typical SME in order that problems during growth can be anticipated and avoided. Businesses generally have readily definable phases of growth during start-up. The business originates from an idea which must be proven. Planning and development then occur, to the point where the new business is ready to start (*Figure 6.4*).

"Remember that ideas always come first – some ideas become experiments, some experiments become projects and some projects become companies."
Oli Barrett, Portfolio Entrepreneur, The Rainmakers

As the business begins to trade it will hopefully enter a phase of expansion/growth. If it continues to expand it may eventually reach a point where the entrepreneur will decide it should continue as a 'lifestyle' business allowing the entrepreneur to achieve the lifestyle he or she seeks with no further need for expansion. Alternatively, the business may be sold to a larger company. A large-scale change of this nature is termed an 'exit strategy' or 'exit route' (see also *Chapters 7* and *8*). If appropriate, it enables the entrepreneur/entrepreneurial team and investors, for example venture capitalists, to withdraw from the company having taken the business as far as they wish to go.

Figure 6.4: Phases of business growth. ▸

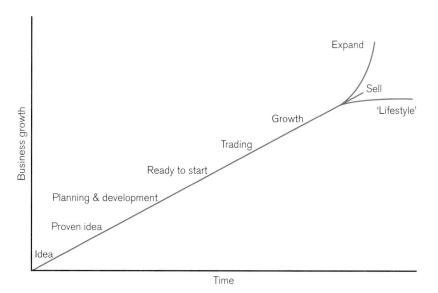

The pre-start-up phase of proving the idea, and then planning and development is particularly crucial. Many 'lifestyle' and non-science based entrepreneurs, or those who are not necessarily interested in business growth, are often able to test their ideas and carry out a great deal of preliminary work long before the launch of the business. For example, they may choose to try selling their product on eBay before establishing a more formal company structure and buying premises (such as a shop). This option of starting small and building up the business is not always open to laboratory bioscientists. Nonetheless, bioscientists should take every opportunity to test their idea through market research that involves assessment of customer demand and the threat presented by products from competitors (see *Chapter 4*).

The UK government has established a series of schemes that finance this phase of start-up. The programmes are funded by the research councils, local centres for enterprise, and other organisations. They can provide a great deal of valuable flexibility for university researchers keen to commercialise the results of their research. For example, the BBSRC provides one-year enterprise fellowships that enable researchers to remain within the university with full access to an external network of mentors, business experts and other professional advisers during the development of the business plan. Other programmes provide help with licensing or some form of seed or venture finance to enable the commercial potential of results/ideas, arising from research council-supported research, to be realised. Further details of these and other funding schemes are provided in *Chapter 8.*

6.7 – Cost implications of starting a bio-business

The Department of Trade and Industry indicates that "the UK is one of the best places in the world for bioscience. The UK boasts an excellent science base, with favourable economic and political conditions, and a clear and fair regulatory regime" (www.dti.gov.uk). However, start-up costs for a bio-business can be particularly formidable with requirements for labs, chemicals, equipment, specialised staff, and the expense involved in attracting these individuals. There are major health and safety, and regulatory issues in the biosciences (see *Chapter 9*) and, as noted by Friedman[2] "a characteristic that sets biotechnology apart from other industries is that biotechnology products must pass rigorous assessment to verify their safety and, in the case of drugs, efficacy prior to receiving marketing clearance". All of these costs must be considered carefully when compiling the financial section of the business plan and these issues will be dealt with in the next chapter.

"The bioscience industry is driven by entrepreneurial companies"

Ian Hughes, HEA Bioscience

6.8 – References

1. **Lang J** (2006) Building a team. In: *Starting a Technology Company,* p. 13–16, Weston-Smith M and Luebcke P (eds). Cambridge Enterprise, Cambridge.
2. **Friedman Y** (2006) *Building Biotechnology,* 2nd edition. ThinkBiotech, Amherst.
3. **Southern M and West C** (2002) *The Beermat Entrepreneur,* Pearson, Harlow.
4. **Belbin RM** (1981) *Management Teams: Why they succeed or Fail,* Heinemann, Oxford.
5. **Nelson L** (2006) In: *Starting a Technology Company,* p. 21–22, Weston-Smith M and Luebcke P (eds). Cambridge Enterprise, Cambridge.
6. **Johnson G and Scholes K** (1998) *Exploring Corporate Strategy,* 5th edition, Prentice Hall, New York.
7. **Hughes I** (2005) Being entrepreneurial – more difficult for some? *Centre for Bioscience Bulletin Spring,* 2005; 7.
8. **Burns P** (2001) *Entrepreneurship and Small Business,* Palgrave, London.
9. **Parks S** (2004) *Start Your Business – Week by Week,* Pearson, London.

6.9 – Additional resources

www.acb.org.uk

The Association of Clinical Biochemists (ACB) is a professional body dedicated to the practice and promotion of clinical science.

www.bbsrc.ac.uk/biobusiness_guide
This site has a range of useful information and several bio case studies to illustrate how academics have created businesses and developed their careers. Drawing upon famous bio-entrepreneurs to illustrate the guide, assists in developing understanding of the start-up approach.

www.bioindustry.org
The BioIndustry Association is the trade association for innovative enterprises in the UK's bioscience sector.

www.biospace.com
BioSpace is a specialized provider of web-based products and information services to the life sciences.

www.the-body-shop.com
This website provides an update on the social dimensions of trading from one of the key proponents in the UK – the Body Shop. As a well known high-street favourite, trading in cosmetics, and defending a 'no animal testing' policy, it provides a useful case and barometer of the changing concerns of the consumer.

www.co-operative-bank.co.uk
Provides further detail to the social aspects of trading, as the Co-operative Bank operates with an ethical banking policy, which responds to changing demands of its consumers, through a regular questionnaire.

www.enterprise.cam.ac.uk/building/starting.html
This excellent booklet provides useful technology start up advice from a UK university. Drawing upon a range of expertise, it provides information on how to start off, getting going and keeping going – including a useful glossary and advice from experts and entrepreneurs that have biotechnology experience.

www.ibms.org
The Institute of Biomedical Science is the professional body for biomedical scientists in all fields of work, including medical laboratory and scientific officers in the National Health Service and related services in the United Kingdom and Ireland.

www.iob.org
The Institute of Biology is the professional body for UK biologists.

www.ncge.org.uk
The National Council for Graduate Entrepreneurship (NCGE) was formed in 2004 with the aim of raising the profile of entrepreneurship and the option of starting your own business as a career choice amongst students and graduates. By understanding the circumstances in which graduate entrepreneurship can flourish, NCGE's goal is to influence, and inspire, an increase in the number of students and graduates who give serious thought to self-employment or business start-up. The NGCE offer a range of opportunities for those at university to experience enterprise, including range of specific regional events – their 'flying start programme'.

www.socgenmicrobiol.org.uk
The Society of General Microbiology's objective is to advance the art and science of microbiology.

www.soyouwanna.com
SoYouWanna.com claims to teach "you how to do all the things nobody taught you in school" and includes advice on business plans, together with other start-up advice, from its useful searchable engine.

Learning outcomes

Key learning points from this chapter are:

- the key elements (skills; knowledge; experience; personal style) that should be considered in the development of a management team to ensure effective start-up and continued growth of any new bio-business

- issues and potential problems associated with various start-up stakeholders, from changing social attitudes through to appreciating the importance of key business relationships

- the forms of business that may be adopted by start-up companies (most commonly limited companies)

- and the alternative strategies to start-up that are available to bioscience and other entrepreneurs, such as spin-out and licensing

- phases of business growth and the opportunity for entrepreneurs to choose between expansion, 'lifestyle' or 'exit strategy/route' as the business grows and develops

- cost implications of starting a bio-business which range from fixed costs such as monthly out-goings (e.g. rent and salaries) and those that increase with success (raw materials needed, etc.)

Chapter **7**
The Role of the Business Plan

Andrew Ferguson

7.1 — Introduction

At the risk of stating the obvious, a business plan is a plan for how a specific business activity will be managed over a set period of time. The business activity in question could be an entire multi-national corporation, a university department, a tiny local charity, or a start-up business led by an entrepreneur. It could be dealing with a turnover of millions of pounds over many years, or a few thousand pounds over a number of months. For the purposes of this chapter, the business plan is assumed to relate to the launch of a new bio-business venture; although, even with this narrowed definition, the plan could relate to a completely new company or a sub-division of a larger organisation. Business plans of this nature tend to consider a planning time frame of up to 5 years with a particular focus on the first 12–24 months. One peculiarity of bio-businesses is that the scientific developments they rely on can take significantly more than 5 years to become commercially viable – the issues generated by this peculiarity will be covered later.

In recent years business plans have assumed great importance in educating managers and potential entrepreneurs. Books, computer programs, and many organisations and agencies offer assistance in writing a 'winning plan' that might not only attract investors, but may also be entered into the growing number of business plan competitions run by universities, banks and regional development agencies alike. However, in all these cases the key to the success of any business plan is clarity of purpose, that is, the author must understand precisely why the plan is being written in the first place.

7.2 — It's good to plan

At a fundamental level, planning is generally a good thing. There are oft-repeated maxims to support this, the most over-used being 'failing to plan is planning to fail'. A business plan can therefore be seen as a genuine attempt on the part of the entrepreneur to set out his or her strategy for the future development of a venture and to estimate the likely implications and outcomes of that strategy. The discipline of preparing a business plan enables the author to draw together all the aspects of the potential company into what is, hopefully, the coherent narrative of a successful venture. For any new enterprise there is tremendous value in the entrepreneur taking the time to pull their thoughts and aspirations together into a single document which sets out the intentions and reasoning behind the proposals. Indeed, if the entrepreneur cannot get the business to 'work' on paper then he or she needs to ask whether they feel confident enough to do so in practice. In the case of an entrepreneurial team, writing the business plan can also ensure that all the team members actually share the same vision of what they are seeking to achieve and how they intend to achieve it. Again, if the team cannot agree on the theory, the practice is likely to be beyond them.

A good plan can also act as a guide for future development, enabling the managers of the business to check on their progress against their predictions and ensure that things are progressing as anticipated or, as is more likely the case, identify why things have varied from what was expected. The German General Von Moltke said that 'no plan survives contact with the enemy' and, as no business plan is likely to survive contact with reality, most entrepreneurs tend to view their plans as 'working documents' to be constantly updated and altered in response to the wide range of variable factors that constantly interact with their business, as described by the entrepreneur in *Case study 7.1*.

Case study 7.1: Medcom

Medcom is a business that provides on-line training in surgical skills. Medcom's product combines advanced IT, training methodologies and in-depth medical knowledge. The company was started and is managed by two entrepreneurs, Chris Matson and Warren Hobben. Their business plan won the top prize from the judges at Venturefest Yorkshire 2003, the region's main event for high technology start-up companies.

However, as Chris Matson explains about their winning plan: "That was probably plan number 38. You've constantly got to keep working on the document because you just cannot predict everything. We learned a lot by just writing the plan and in having to constantly update our thinking by spotting the holes in what we were intending to do."

In reality, a great many SMEs (small and medium sized enterprises) actually have no operational business plan at all. The day-to-day pressures of running a business often make the process of looking beyond the next customer's order a luxury that only a well-resourced and well-staffed organisation could hope to afford. The concept of the business plan was not developed by entrepreneurs; indeed the skills and aptitudes of the planner and the entrepreneur are rarely found in the same individual. The business plan was developed instead as a requirement of those who finance enterprise: banks, venture capitalists and business angels, and most plans are written with these investors in mind.

7.3 – The business plan as an offer

If an entrepreneur is convinced that a plan can work in practice, then the next stage will be to win the resources to invest in the means to establish and run the venture. The great majority of businesses in the UK draw on the resources of the entrepreneur and his or her family and friends to do this at the start; even EasyJet was started with family money, albeit in a family with millions to invest. For bio-businesses, especially those involved in pharmaceuticals development, the scale of the resources required means that family and friends simply are not an option. Some forms of bio-business can 'burn' their way through millions of pounds without even getting close to putting a product into the marketplace. This means that the entrepreneur must convince investors of the attractiveness of the potential business. In these circumstances the business plan becomes a detailed sales document, setting out an offer of an anticipated reward in return for an investment.

A business plan used for this purpose is still the narrative of how an enterprise will become successful, but lays far greater emphasis on the needs of a reader from outside the immediate entrepreneurial team. In developing a business plan to attract investment from any source the author has at all times to consider what the reader needs to know. Most will want to see clearly and realistically how the business will become commercially viable and what potential rewards their investment will bring them. However, the author should also be aware of the slightly differing requirements of the distinct classes of investor.

There are a number of potential sources for investment in bio-business, each with their own peculiar features in terms of the rewards they are likely to be seeking from investment in a new business. The author of the business plan therefore has to structure and tailor their plan to meet the interests and informational requirements of the reader. As with any sales document, the business plan must understand the benefits the reader is looking for and ensure that not only are these contained within the plan, but also that they are easy to find and understand. *Chapter 8* provides an overview of the common sources of investment to which a start-up bio-entrepreneur might turn, but these are previewed here to show how a business plan might vary in emphasis depending upon which source is being approached.

7.3.1 Banks

Most commonly banks offer loan finance. A bank will obviously want to charge interest on their loan but, subject to possible variations in the interest rate, this is a fairly predictable level of on-going commitment for the business. For an entrepreneur the attraction of a loan is that the bank is unlikely to want any share in the ownership or management of the business. Whilst this might appeal to an entrepreneur who values independence, it may also be a lost opportunity to bring more experienced individuals into the venture. Although some banks have made efforts to develop in-house expertise in the biosciences, they are generally faced with a wide range of businesses seeking support and have something of a reputation for being extremely cautious in lending to this sector. A business plan for a bank will need to be very clear on the commercial value of the technology involved and to emphasise the stability of the enterprise and the security and regularity of the revenues that will allow it to meet its repayment commitments.

7.3.2 Business angels

These are often individuals or small groups who have made a success of their own businesses and now have the time and money to invest in new ventures. Angel investments form a midrange between the operational funds lent by banks and the really big sums handled by venture capital firms. As they are usually investing their own money, angels are likely to want to know about the team behind the plan and how they can interact with them. Typically, angels are interested in new ventures where they feel they have some form of experience or expertise to contribute. A return on investment is, of course, important to a business angel, but so is the interest and excitement generated by contributing experience and expertise during the launch of a new venture. This means that they will often look for a significant 'share' of the business in order to exert their influence.

7.3.3 Venture capital

The scale of investment required by some bio-businesses means that venture capital is commonly associated with start-up funding in this sector. Venture capital tends to operate in terms of millions of pounds rather than thousands, although there are some specialist funds that will consider smaller investments. Venture capital is used to 'buy' a share in the new business activity, thus providing it with the funds to start up and grow.

Unlike a loan, which has to be repaid with interest regardless of the success of the business, venture capitalists get no return if the business does not grow in value. In return for this risk they hope to recoup their investment many times over by eventually selling their share to another business, or on the stock market, for far more than they paid for it. Put rather simplistically; they hope to do this in as short a time as possible for the maximum value. As a general rule, the greater the risks involved in a business, the higher the rate of return a venture capital investor will expect.

This means that a business plan aimed at attracting venture capital has to stress the growth potential of the business and make it clear when and how the venture capitalist can look forward to getting their investment back again, together with a commonly anticipated additional return in excess of 40%. How the entrepreneur expects the investor to realise this return is called the 'exit route' (see also *Chapters 6* and *8*). In the case of some pharmaceutical businesses, where development can take 10–15 years, the issue of exit routes can be particularly problematic.

Other than the money that venture capital companies can bring to a start-up there is also the potential to bring in proven experience in terms of strategic management. Specialist bioscience investors will generally have access to individuals who have themselves been entrepreneurs in the sector. This is known as 'smart money' and is well illustrated by Merlin Biosciences, a UK-based specialised venture investor in life sciences (see www.merlin-biosciences.com). In their business description they are clear that they will want to be involved in the direction of the ventures in which they invest: "Consistent with its entrepreneurial and scientific approach, Merlin is an active investor in and partner to its portfolio companies. Merlin normally has a seat on each company board and likes to play a leading role in financing and other strategic events."

Again, a business plan will need to reflect a willingness on the part of the entrepreneur to welcome this type of involvement. Indeed, it is likely that a well-structured business plan will include reference to the inclusion of some expertise known and trusted by potential venture capital investors.

7.3.4   Grants and special awards

Along with digital technology, the biosciences are seen as key to the development of what has become known as the 'knowledge-based economy'. This means that just about all organisations concerned with economic development in the UK are keen to see bioscience businesses setting up in their backyard. Grants and special schemes can be found to help this happen and could, potentially, be very useful to a start-up business. Often the application processes for these awards is, or closely resembles, the development of a business plan.

However, there is very rarely a 'free lunch' to be had in this area. All grant-awarding programmes have strict criteria for judging potential applicants and applications. The entrepreneur will need to be familiar with these criteria and structure the business plan around them. He or she will also need to assess whether the benefit of the grant support outweighs the possible restrictions of the criteria. For example, a grant tied to starting up a business in a specific geographical location may not be appropriate if the expert staff the business is likely to need are known to be clustered at the other end of the country.

7.4 – Drafting a business plan

In general terms, a business plan should be as easy to read as possible. A business plan is trying to convince a reader that a business idea is not only viable but also exciting and worthy of commitment. In the same way that a novel uses a narrative structure to carry a reader through a story to a satisfying conclusion, a business plan is trying to convince the reader that their investment is the logical and convincing conclusion of the document. This means avoiding repetition of unimportant information and pushing important but supporting information into appendices at the end where it won't break up the logical flow of the plan.

Whilst the plan is a form of sales document, it must be careful not to over-hype or misrepresent what the business is seeking to achieve. In the biosciences in particular, where specialist investors are likely to have a well-developed knowledge of the sector, they will expect there to be risks involved and will be looking for a business plan that convincingly identifies potential pitfalls and inspires confidence that they will be managed effectively.

As with all documents selling to an audience, from CVs to sales letters, it is important for a plan to create a good initial impression. This means taking some obvious steps in terms of presentation:

- text should be typed and it is recommended that diagrams and financial data are produced using appropriate software rather than being drawn by hand
- putting a good-sized margin on the page not only helps avoid threatening the reader with too much text but also allows an investor the opportunity to record their own notes
- plastic covers on a spiral bound plan help protect it from wear and tear in the in-tray
- the cover sheet should state the name of the business and have contact details for the author — the plan is very likely to become separated from any cover letter
- numbering the copies of the plan allows the authors to keep track of who they have been sent to and, provided that the number of copies is limited, is also an assurance to the reader that he or she is being included in an exclusive group of readers
- it is important to avoid the use of technical jargon if at all possible (see also *Chapter 5* for tips about how to communicate science to non-scientists)

7.5 — Structuring the plan

Whilst there are many variations on exactly how a business plan should be set out, there are certain conventions as to what an experienced reader of these documents will expect to see. *Table 7.1* shows a potential business plan structure.

Table 7.1: A business plan structure

Section	Summary description
Cover page	Gives the title of the enterprise and, importantly, the contact details of the author
List of contents	Helps the reader find what they are looking for quickly and logically
Executive summary	The business plan on one side of an A4 page. If the business has a particular view of its vision, mission or approach it could be included here
The people	Identifies the management team of the business and establishes their credibility in putting forward a business plan in this area
Background	Sets out where the business idea came from, and the marketplace and conditions in which the enterprise will operate, for a reader who may be unfamiliar with this information
The product or service	What is it the business will do? How will the product or service differ from those already available? Is there any IP involved?
The marketplace	Who will buy the product or service? How many customers are there? Why will they buy from the business and in what quantities and for how long? Are there any wider contextual issues which will be working in the business's favour?
Competition	Who else might they buy from? Why will they buy from the proposed business instead?
Strategy and plans	Sets out the business's objectives and details, and exactly how these will be achieved. Why are the skills of the individuals involved ideal for making this happen?
Financial projections	How does the analysis of the market translate into sales and revenue? What assumptions underlie these projections? How will the business bring in income and incur costs over time?
The investment proposition	What investment is required? What share of the business is on offer? What is the exit route and what return will the investor receive?
Risk analysis	What factors are critical to the success or failure of this venture? What can the management team do about these factors?
Conclusion	A final message to leave with the reader
Appendices	Supporting information that is important but would detract from the flow of the main document; might include: CVs of the management team, market research data, letters of commitment or support from customers

7.5.1　Executive summary

The summary is probably the most important part of the entire plan. Faced with a number of business propositions the busy investor will turn immediately to the summary. If the first few lines don't grab the investor's attention then the chances are that all the carefully researched and clearly set out text in the following sections will go to waste. This means that knowing your audience and including their particular interests in the summary is especially important.

The summary is not an introduction. It is the business plan condensed down to less than one side of A4. This can be difficult, but is never impossible, even for the most technically involved business. A summary needs to include:

- the product or service at the heart of the business and why this is special, not in terms of how it does what it does, but in terms of what benefits it will have for a customer
- who the entrepreneur(s) is and why they are particularly suited to make the business succeed
- the marketplace in which the business will operate and the advantages it will enjoy there
- additional advantages the business has, e.g. any patent it holds
- what the business will do to achieve success
- what financial value the business will generate
- what is asked of the reader, i.e. the level of investment required

The inclusion of some headline figures in this section is important. If the reader is an investor they will need to know how big a proposition the plan represents. For example, an indication of the financial value of the business's annual target market, and the percentage market share anticipated, gives a sense of scale to the plan.

Some elements of the executive summary are best written before the rest of the plan and some afterwards. It is important to revisit the executive summary at the end of writing the whole plan to ensure that the conclusions of the detailed plan concur with those of the summary.

If those involved in the business feel it has a particular mission, vision or philosophy they should also state it in the summary. These concepts can help shape the rationale for the business:

- vision – where the business sees itself going in the future
- mission – what will it try to achieve right now
- philosophy – how will it interact with the world along the way

A vision is one of the main reasons for writing a business plan in the first place. It is what will carry the business forward and should be simple enough to be expressed in a single sentence: 'To be Europe's most profitable supplier of bio-remediation materials' or 'To have the largest customer base in our field' would be simple statements of vision. The mission should be as precise as possible, usefully over the next two years, and be based on customer needs and benefits. It should include clearly identifiable goals, such as sales figures or profitability; for example, 'to grow turnover to £1 million p.a. by the end of year two'. Philosophies may also be appropriate if these will influence how the business will be run. Famous company philosophies have been those of the Body Shop and Ben and Jerry's ice-cream; both businesses are as famous for their way of doing things as they are for the products they offer.

7.5.2 The people

It is often said that venture capital invests in people rather than ideas and that a team with a good reputation and an average idea have more chance of attracting funds than an excellent idea backed by a team of unknowns. There is a fair amount of truth to this, as management ability, technical skills and previous experience are often the deciding factor in taking an idea from exciting concept to commercial reality. Establishing the credibility of the team involved from the outset is therefore to be recommended (see also *Chapter 6* regarding the importance of team building when starting up a business).

As with general recruitment and selection, the reader is looking for evidence from the team's past experiences and achievements that can give an indication as to their likely success in this particular venture. In a plan for a start-up business it is unlikely that those involved will have experience of a business in exactly that context. It may therefore be necessary to pick out achievements that are as relevant as possible. For example, a sales director who has already had success in finding customers for a new bioscience product has some track record to suggest that they can do it again within the same sector.

At the business plan stage it is important for the entrepreneur to recognise and accept that there may be gaps in their own skills which could trouble investors. Gaps can, and should, be filled by recruitment at the planning stage. Indeed most investors will insist on seeing some experienced start-up managers in high value investments, no matter how bright and technically able the rest of the inexperienced team are. Finding good management skills has become something of an issue for bioscience businesses developing out of academic research. Although there are some notable exceptions, those with established experience in biological research in a university context are unlikely to also have exactly the right skill sets to launch and manage a small commercial enterprise. This often means bringing in a manager who has successfully led a bioscience venture in the past. These people are often in short supply and, by virtue of their previous success, can be far more expensive than many smaller start-ups can afford.

The team should be introduced using short (one paragraph) profiles which include some basic information on education and previous employment history. Particular emphasis should be placed on achievements and specific experience within the role identified for the team member in the new venture. The reader should gain the impression that the team has been carefully assembled for their complementary skills, rather than that they all happened to come together as friends who worked in the same lab at university. Full CVs of the team are likely to be too cumbersome for the main body of the business plan but should always be included in an appendix.

7.5.3   Background

This is an optional section which can be used to set the rest of the plan in some kind of context. For example, it might give a short history of how the business idea came to be developed and why the entrepreneur is so convinced of and committed to its success. In some cases the author may feel it necessary to explain a little about the marketplace the plan will be dealing with. Although there are some specialist investors around, it cannot be guaranteed that the reader will have any experience or knowledge of the field the plan will be dealing with. This is especially likely in areas where the technology is new or its application particularly unusual. For example, when Medcom (see *Case study 7.1*) were putting their business plan together they found it useful to include a summary on the particular niche market their service was to be involved in. Chris Matson discovered that "Whilst investors had a good working knowledge of the health care industry, we found that they had less idea of the details of how surgical training and development is handled in the NHS. We included a one page background of the sector in front of the executive summary so that all readers could immediately see how our business was going to fit into that industry."

The background section should also be kept as short as possible, setting the scene for the information that will follow. It should avoid discussion of areas that, although potentially interesting, add little or nothing to the case for investment in the business.

7.5.4   The product or service

This is the section that many business plan authors look forward to the most. It is the part in which the output of many hours of research and development are finally explained to the world. Sadly, most investors are simply not interested in the technical processes that lie behind the product or service (see *Case study 7.2*). Instead, they would rather know:

- what it does
- the benefit to the user
- what is different or unique about it, and why it is better than anything comparable
- whether it is protected in any way, e.g. is it patented?

Whilst it may be difficult to explain exactly what the product is and how it works without the use of detailed scientific terminology the author must be crystal clear on its eventual value to the customer. This also means that the author should know exactly how the potential customer defines 'value'. For example, saying that your product will help horticulturalists grow larger tomatoes is not, in itself, valuable. To yield value to the customer, the grower will need to be able to sell his newly enlarged tomatoes at a sufficient price

Case study 7.2: Nick Butler

Nick Butler is Executive Director of Connect Yorkshire, an organisation that helps technology-based businesses prepare for the investment market. An experienced investor himself, he has dealt with mergers and acquisitions, floatations, and venture capital transactions, and was awarded the Dealmaker of the Decade title by *Insider* magazine in 2001.

He identifies overly long descriptions of the science underlying a product or service as one of the most common errors of technology start-up business plans. "It isn't surprising that the authors concentrate on explaining the science; it is probably what their background has prepared them to do. For the business plan it can be taken as read that the technology will do what the author says it will. Proving that it really works comes later at the due diligence stage."

Nick is also very clear on the importance of a strong market analysis in a business plan; "The first thing I am looking for is the route to market. If possible I would like to see a track record of growing sales already in the business, but the inclusion of letters of support and commitment from future customers for a start-up business is also reassuring."

premium not only to afford to pay for your product but also to make it worth his while changing his production process to include it in the first place. The evidence to back up the statement of customer value comes in the next section of the business plan on the market and market analysis. What should be included here is a 'headline' statement of why the product/service is going to be in demand by the target customers.

IP is often a vital feature of the bioscience business plan. Having a legally enforceable patent on a feature of the product or service considerably reduces risks to the investor and demonstrates very strong commitment to the business on the part of the authors (see *Chapter 3* for full details of this crucial area). Bear in mind that if the business plan is prepared without IP or IP protection already in place, then the investor is likely to demand a greater share of the returns as a reflection of the increased level of risk involved.

7.5.5 The market and market analysis

Along with the people involved, the strength of market demand for the product or service offered is an essential factor in determining whether a business plan can attract investment (see *Case study 7.2*). It is essential that the business plan demonstrates an in-depth and comprehensive knowledge of the marketplace the business will operate in (see *Case study 7.3*). The entrepreneur must therefore be aware of both the customer base and the nature of the competitors. A key element of this is the acquisition of 'competitive intelligence', the form of market analysis that places strong emphasis on the tracking and monitoring of direct competitive behaviour (see *Chapter 4*).

Case study 7.3: BenGay

The US pharmaceutical company BenGay, which is best known for its pain relief creams, attempted to launch a range of orally taken analgesics. Market research indicated that the company had a solid reputation for its creams and sprays and it seemed logical to create a suite of pain relief products trading on this brand. Consumers, however, associated the brand with warming creams, to the extent that the idea of swallowing a BenGay product was both unusual and, for too many, unattractive. Market research only told part of the story; that people trusted the company for pain relief products. It failed to identify that this trust extended only to a particular type of pain relief product.

Customers

With regard to customers, the author of the business plan will need to know the answers to a number of deceptively simple questions:

What problem does the business' product or service solve? Putting the results of the product or service into terms that are immediately obvious is one of the first steps toward attracting an investor's attention. Sometimes called the 'so what?' test, a business plan should be able to state clearly and simply what the product or service will do for customers and why that is a good thing. A business plan offering a product that 'significantly improves the nutrient content of soils used by specialist growers' easily attracts a 'so what?' from the investor. However, if the plan were to rephrase this as 'allows growers to achieve additional yield which will enable the product to pay for itself within six months' the advantages become immediately apparent.

Who currently faces this problem and is therefore likely to be willing and able to pay for the solution offered? A relatively detailed customer profile is an important element of a business plan as it demonstrates that the busi-

Case study 7.4: CueCat

One of the costliest 'why bother?' innovations of recent years was the CueCat personal bar code scanner. This device consisted of a scanner pen with which a PC owner could scan a bar-coded internet address from a magazine into their computer taking them directly to the advertised website. The premise of the business was therefore that people reading a magazine would get up and make their way to a PC to look at a website, the address of which was so obscure that they could neither guess it, remember it or be bothered to copy it out. Consumers responded with a resounding 'why bother?', but not before CueCat had burned through an estimated $180 million of venture capital.

ness has clarity in who it intends to market to and why. However, there are numerous examples of businesses that offered solutions to issues that the intended customer did not actually perceive as problems worth dealing with (see *Case study 7.4*). Faced with any potential change most people will ask '*why bother?*'. They will expect a simple and compelling answer if they are to adopt the product or service, and so will any potential investor.

What advantages will the customer perceive the product or service to have? These might not necessarily be the advantages the author of the business plan initially thinks their brainchild offers. Advantages must be stated clearly and in terms relevant to the customer and understandable to a commercially aware if not technically minded reader. For example, Authentix (see also *Case study 1.1* and www.authentix.com) is a company which uses patented antibodies, developed using advanced biotechnology, to detect adulterations in other products. However, the company makes very little of its technology in its marketing; "Our clients don't have technology needs. They have commercial problems which technology may help solve. In an industry awash with niche technologies we never forget this. Our principal goals are to define the issue, design and install the most appropriate solution, manage the program and most importantly, deliver the commercial return. In the last three years we have saved our clients over $3 billion in lost revenues in the petroleum, pharmaceutical and consumer goods industries alone." A bottom line saving on lost revenues is a concept that investors will grasp immediately.

Why will the customer take up the solution the business offers and not someone else's solution? What is the so called 'unique selling point' or USP of the product or service and how does this help the business pull away from its competitors? Again, the '*so what*' test should be applied rigorously to any USP a business claims. Every business has competitors no matter how innovative or new the technology involved (see below).

How many of these potential customers are there? What scale does the market have? Is it growing, static or declining? This might be expressed in numbers of customers or in the financial value of the market. It is essential to have objective, third party figures on this and to be relatively precise in the definition of the market the business will be seeking to address. The temptation is often to define the market very broadly and assume this gives you a lot of potential customers. For example, a business selling food-stuffs should not assume that everyone in the UK is its target market, despite the fact that everyone eats.

How many potential customers are realistically going to become actual customers, and over what time period do you anticipate this happening? Business plans sometimes assume that because a potential customer *could* buy the product and service they *will*. Business plans notoriously overestimate sales, probably because entrepreneurs are understandably up-beat about what they are seeking to do. Nick Butler (see *Case study 7.2*) is very familiar with the overestimation factor; "the author will try to reduce his estimated sales to appear more realistic. Most investors will take this reduced estimate and at least halve it."

How can the business ensure that the customer will continue to be a customer if this is required for business growth? Winning customers is often only a part of the story for a successful business. It is often easy to assume that once business is won it will be retained, allowing for a nice steady growth curve on a sales chart. In reality, retaining customers requires as much effort as winning them in the first place. What specific strategies will the business adopt to ensure that customers buy again?

Competitors

All businesses have competition. Even in those extremely rare cases when a technology is so new the company can legitimately claim it to be the first of its type, there will still be competition (see *Case study 7.5*). The entrepreneur must demonstrate a thorough understanding of the competitive environment and this can be achieved through the acquisition of competitive intelligence (see *Section 4.3*).

Other factors

A bewildering number of variable factors impact on a business. These range from fluctuations in currency markets to trends in demography, from

Case study 7.5: Amazon

The huge initial investments made by the founders of Amazon were an attempt to build a dominant market share in a brand new market – online book sales. Amazon was largely successful in this, although it took years for the business to return a modest profit. However, although online competitors have largely been mastered, Amazon still faces competition from traditional bookshops, some of which have responded to online rivals by making a distinct virtue of being able to browse and touch their goods in a relaxed and non-virtual way through in-store coffee shops and armchair seating.

Medcom, the online surgical skills business described in *Case study 7.1*, occupies a similar market-leading position. However, while it offers a unique means of surgical training it still competes with decades of established medical school practice. As Chris Matson explains, "We took a very broad view of our competition in our business plan; we examined textbooks, clinical practice and class room teaching and looked at the strengths and weaknesses of all those in relation to our own product. We presented this analysis in a table, being as precise as possible about whom the competitors were, actually naming them if possible."

legislative initiatives to raw material prices. Most of these factors are beyond the control of governments and multi-nationals let alone start-up businesses. Nonetheless, a plan should address any contextual issues that are likely to be working in the business's favour in the market section. For example, in recent years there has been a marked increase in the number of individuals aged 60 years and over in the western world. At the moment, this aging population is relatively affluent; again objective evidence is available to support this assertion. Therefore, a business planning to sell products that will improve the quality of life for the elderly can point to both a growing demand for their product along with consumers with the resources to pay for it.

Legislative frameworks also impact heavily on the bio-business sector. Regulations add time and cost to the development process of many products in the sector, albeit to protect the end-users. Whilst it is always worth accounting for the impact of regulation on the costs and activities of a new business at the planning stage, it is also worth considering whether regulation offers any competitive advantages (see *Chapter 9*). For example, Xceleron – a university spin-out, introduced Accelerator Mass Spectrometry (AMS) into critical stages of the drug development process, helping pharmaceutical and biotechnology companies improve drug candidate selection, reducing the risk of clinical failure, and streamlining both the IND (investigational new drug) and clinical development processes. The AMS technology has been used to assist over 75 biotech and pharmaceutical organisations, including 15 of the world's top 20 pharma companies, in their drug development activities. Xceleron offers its clients a reduction in attrition at later stage testing and therefore saves valuable time and significantly reduces downstream costs. In effect, Xceleron uses the regulation and testing regime as the framework for its market.

Prove it …

Investors reading a business plan are looking for impartial and objective information to support what the author is telling them. Telling an investor what you believe is nowhere near as powerful as providing verifiable evidence to support your claim. A plan that asserts that there is a growing market for a product because the author has received a 'lot of interest' from those in the industry looks weak against a plan that can quote and reference a market growth figure from a recent *Financial Times* survey. A further point is that expressions of interest from customers prior to the establishment of the business will be very attractive to investors. Letters of support from future customers, or even getting customers to put resources into the development process of a new technology they are keen to use, is evidence that the business plan really understands its 'route to market'.

Venture capitalists the world over are still recovering from one of business history's greatest triumphs of unfounded optimism over the sober research of marketing basics – the dotcom boom and bust of the late 1990s and early 2000s. In the UK alone, the top 10 internet company failures between them raised and wasted £520 million of investment. Part of the reason for the spectacular failure of these businesses was the perceived lack of any precedent business models on which to base research and due diligence. It was 'all new, all revolutionary' and so investors had better get on board or miss the opportunity.

Dotcom investment was also fuelled by misleading market information. One such piece of information was that 'internet traffic was doubling every 100 days'. This information was quoted in a US Department of Commerce document during 1998 and in numerous investment pitches; despite the fact it was later proved to be simply wrong – internet traffic was doubling, but every 12 months, not every three.

Since the dotcom crash in the early 2000s, investors have been particularly keen to see a variety of reputable market information sources quoted in business plans. The claim that the idea or technology is so new that no precedents exist is unlikely to convince them again for a few years to come.

7.6 – The operational plan

So far the business plan has presented:

- core objectives, presented in an Executive Summary – possibly including a vision or mission statement for the business
- the competencies of the team members
- the market context in which the business will operate, including a clear picture of who the customers are and why they will be buying
- who the competitors are and why the new business will be superior in the chosen marketplace

Next the author must draw these strands together into an operational plan. The factors already considered should enable the author to develop a plan that plays to the competitive advantages of the business, whilst also recognising the realities of how the chosen market works. As shown in *Figure 7.1*, the first step is to give the new business clear objectives.

Objectives need to be defined clearly as these will be used to judge whether or not the business's subsequent strategies are working. Objectives need to be specific and preferably involve measurable criteria, such as levels of turnover or profit, and have a timescale against which they will be achieved. Such quantifiable indicators of success, set out against a realistic timescale, give the business plan 'milestones', or points at which

Figure 7.1: An
operational plan. ▸

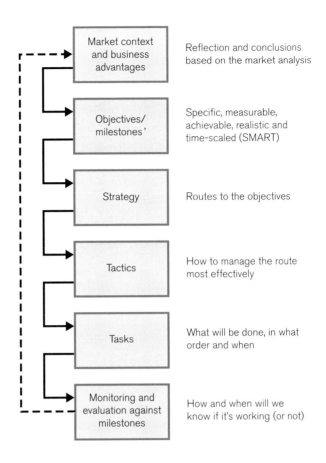

Figure 7.1: An operational plan. ▸

Market context and business advantages	Reflection and conclusions based on the market analysis
Objectives/milestones'	Specific, measurable, achievable, realistic and time-scaled (SMART)
Strategy	Routes to the objectives
Tactics	How to manage the route most effectively
Tasks	What will be done, in what order and when
Monitoring and evaluation against milestones	How and when will we know if it's working (or not)

progress can be assessed and measured as the business develops. Clear milestones give the reassurance that the business will have direction and a management that knows where the company is heading.

The next stages are to develop strategies and then tactics to meet the objectives. Essentially, strategies will offer a route to where the business wants to be, as defined by its objectives. A strategy doesn't deal with the practicalities of taking that route; these are the tactics the business will use along the way. For example, saying that a business will become the UK's leading supplier of a product in terms of annual turnover is an objective. Then saying that it will achieve this by sourcing the product from low cost manufacturers overseas and selling it to UK users at a rate consistently cheaper than the competition is a strategy. The plan then needs to go on to say how the product will be promoted and distributed to UK users and how low cost deals will be negotiated with the manufacturers; these are the tactics that will need to be used.

Strategy can cover a number of areas and will depend on the type of business and, crucially, the market in which it will be operating. A single business should have a single strategy although this might be built up from a number of component strategies.

- *Research and development.* This is sometimes the driver for start-up bio-businesses where the whole strategy is focused on developing a product with strong patent protection. Once the product is known to be effective, the business may sell or license it to a larger company and can then either disband, with the entrepreneurs and investors considerably richer, or move on to the next R&D project.
- *Production.* Usually aimed at getting a lower cost or higher quality product into the marketplace.
- *Marketing and sales.* This strategy aims to create a demand for the company's activities. The strategy might incorporate promotion, branding, routes to market, or price. Price is an element of the strategy that will be carried over into the business plan's financial forecasts. It is not always an advantage, or necessary, to be the 'cheapest' in the market. Pricing for a sustainable business is very much a matter of market awareness and long term strategy.
- *Other elements.* Other factors that might be strategic drivers are location, legal aspects, financial management, technology use, human resources, customer service provision.

With a clear overall strategy, the operating plan (the tactics to be used) to achieve the strategy can now be set out in the business plan. Tactics generate a whole range of tasks to be achieved against a timescale. A lot of time and effort can be saved, both for the author and the reader of the business plan, if the operational plan uses one of the established project management presentation techniques to set out these tasks in a summary format. Such techniques include:

- Gantt charts
- PERT (programme evaluation review technique)
- network diagrams

More elaborate project management software might also be useful in generating an operational plan, although the temptation to produce an unnecessarily complicated plan, using all the software features, should be resisted. Sometimes just using a pad of 'Post-it' notes to sort the operational steps into a sequence, and producing a basic Gantt chart of the results, may be more appropriate. *Figure 7.2* shows the type of Gantt chart that might be used when planning a business's sales activities during its first 12 months. Chris Matson of Medcom (see *Case study 7.1*) found that "putting our production plan onto a Gantt chart probably saved us about four sides of text".

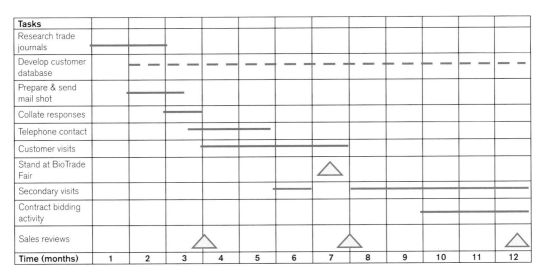

Tasks												
Research trade journals												
Develop customer database												
Prepare & send mail shot												
Collate responses												
Telephone contact												
Customer visits												
Stand at BioTrade Fair												
Secondary visits												
Contract bidding activity												
Sales reviews												
Time (months)	1	2	3	4	5	6	7	8	9	10	11	12

Figure 7.2: Gantt chart for a business's sales activities during its first 12 months. ▲

7.7 – Financial projections

The financial projections should be clearly derived from all that has gone before. The financial part of a business plan is sometimes perceived as the most important and accurate element of the plan. However, it is always worth recalling that financial projections are, at the end of the day, someone's best guess. Of course that guess will have been crucially informed by a detailed market analysis and operational plan but the plan remains, at best, an estimate.

With some variations and additions there are essentially three vital pieces of financial information required in most plans:

- sales forecast
- profit and loss account
- cashflow forecast

Sometimes included in this list is the balance sheet but this can be of limited value for a start-up business as it shows a snapshot of finances at one moment in time. As a result it depends on what has gone before in terms of sales, income, expenditure, and other company operations. In the case of a start-up nothing has gone before so a project balance sheet will have to be derived from the other three forms of information.

7.7.1   Modelling the 'what ifs'

All business plans are built on a number of assumptions and for a plan to show the financial implications of them all would be burdensome to both the author and the reader. The author therefore needs to show that he or she has identified the crucial determinants of success for the business and should demonstrate a limited number of the fundamental scenarios in the financial information.

The spreadsheet is the essential tool for financial forecasting from sales through to cashflow. Readily available spreadsheet templates are useful in the financial section to help ensure that the plan is using presentation formats that conform to those commonly understood by investors. A number of websites offer downloadable formats and many banks will provide template software free of charge.

By setting up a spreadsheet 'book' it is possible to get information to read across all the individual forecasts, from sales through to cashflow. By doing this the author can 'model' a range of scenarios, immediately seeing the implications of variations in forecasting assumptions. The process of considering the financial implications of how the major 'what ifs' will impact on a business is known as *sensitivity analysis*. The author can demonstrate a grasp of the risks involved in the enterprise by presenting a range of financial forecasts based on optimistic, realistic and downright pessimistic scenarios.

7.7.2 Sales forecast

Whilst not all the subsequent financial projections are generated by the sales forecast, most of them are. The sales forecast helps to start the following process:

1. calculation of how the marketing strategy will generate sales; how many and over what time frame
2. having first decided on a clear pricing strategy, the sales forecast can then be used to generate the anticipated amount of revenue
3. the costs of funding the activity required to meet the forecast sales will generate a figure for the cost of sales
4. the business plan can then present an estimate for gross profit (sales less costs)

Forecasting sales depends on a whole series of assumptions. These might include macro-economic issues, such as interest rates, or largely unpredictable features such as the weather; they may be relevant only to the individual business, for example, the securing of a particular contract or getting a patent in place by a certain time. Realistic assumptions, informed by the market analysis and strategy set out previously in the plan, are perfectly acceptable to an investor provided that they are clearly detailed alongside the forecast. The plan should be building an explanation for the sales forecast and should not merely assert the numbers.

Unless there is very good evidence to suggest otherwise it is usually unwise to take the 'proportionate' approach to sales forecasting. This approach establishes the total size of a particular market (say £500 million p.a.) and then says "We conservatively project that we will capture 1% of this market in 12 months, giving us a turnover of £5 million at the end of year 1". Investors are rightly sceptical of this blunt form of forecasting and are very wary of the word 'conservatively'.

Investors welcome cold realism in the development of sales forecasts. For example, it is unlikely that a business can go from a standing start to near optimal sales in a matter of weeks. It will take time for sales staff to contact potential customers, take time to get through formal procurement procedures and time to actually negotiate contracts. Similarly, it is unlikely that sales will move ever upward, even within the relatively limited timescale of a business plan. Factors such as product life-cycle, changing competitor behaviour and new technology can and will impact at some point.

7.7.3   Profit and loss account

No commercial organisation can last for long without a profit and no start-up will attract investment unless profit can be predicted at some point. The profit and loss account (P&L) is therefore a relatively straightforward, although not wholly reliable, measure of a business's success. To find the profit of a business means understanding, and being able to detail, not only its income but also its costs. Costs come in all shapes and sizes and can be predicted, and subsequently controlled, by the entrepreneur with a far greater degree of precision than sales. A spreadsheet of the business's predicted costs should form part of the financial forecast.

The generation of gross sales revenue will have a financial cost to the business, and this is known as the 'cost of sales'. An entrepreneur needs to know not only what the market will pay for a product or service but also how much it will cost to produce each unit. This is true of a single pharmaceutical tablet, an hour of consultancy advice, or a tonne of fertilizer.

Some of the costs of getting the product to the customer are 'direct'. These can, very simply, be divided into two broad groups:

- fixed costs – those which won't vary, whether the business is extremely busy or stands idle, such as rent of premises
- variable costs – those which will vary with the levels of business undertaken, such as raw materials and distribution

However, there are other costs which are 'indirect', sometimes known as operating costs, which also need to be accounted for. These costs can include:

Depreciation: buying a capital item, i.e. an item with a productive life of more than 12 months, such as a piece of high technology equipment for production purposes, is often vital for a bio-business. Showing the full cost of this item on the P&L in year one of the business's activities would make it look very unhealthy, but in year two there would be a massive reduction in costs. Depreciation smoothes this process by writing off the cost of the item over its anticipated working life to give a clearer impression of how the business is really performing. For example, an item worth £50 000 with an anticipated useful life of 5 years is shown in each year's P&L as costing the business £10 000 in depreciation.

Employee costs: all staff activities should be contributing to the running of the business and are therefore P&L costs.

Marketing and promotion: it is difficult to separate out the 'bit' of marketing that leads directly to a sale, because marketing is a complex mix of direct promotion and awareness raising. Unfortunately, all of it tends to cost something.

Research and development: in some bio-businesses this is the focus of nearly all activity. There may eventually be only one actual sale, that of the output of the R&D to another business that has the capacity to manufacture, market and sell the results.

Operating costs offer the manager a degree of flexibility in planning a business. Deciding how much should be spent on promotion and marketing, and how much on R&D, are questions of fine judgement and may be open to question by experienced investors reading the plan.

7.7.4 Cashflow forecast

It is sometimes said that accountants produce profits whilst businesses produce cash. It is certainly true that a business goes bust when it hasn't got the cash to pay its creditors, but it need not necessarily go bust when it makes an accounting loss. Cashflow is therefore vital to the survival and success of the business and is, in many ways, a better indicator of viability than the P&L account. Whilst major capital expenditure is spread over the life of the equipment on the P&L, the impact of such a purchase on the business's cashflow is immediate and potentially fatal if the funds are not there to meet the cost. Forecasting cashflow is also the means by which the entrepreneur can gauge how much investment is needed to get the business up and running.

Cashflow is derived from the sales forecast and a clear understanding of the likely costs the business will incur to meet those sales. It also depends on understanding how purchasing and sales are handled in the market the business will operate in. For example, a business that simply trades chemicals rather than producing them will need to understand what credit terms it can obtain from its suppliers and what terms it can safely offer its customers. Can customers be expected to pay 'upfront' for their chemicals or by using staged payments? What form of payment terms do the competitors offer and do these make them more attractive? The entrepreneur must beware of payment terms because an apparently profitable contract can bankrupt a business if payment arrives too late to cover the costs of delivering it.

Figure 7.3 shows graphs of the cashflow for two different kinds of business. Both lines are artificially smooth for presentation purposes, because in reality cash inflow and outflow will peak and trough month-on-month. Line A represents a subscription-based specialist newsletter on research funding and line B a bio-business researching a bio-remediation enzyme for dealing with contaminated land. The set-up and running costs of the newsletter are relatively modest: computers, payments to contributors, some promotional expenditure. However, the market is relatively competitive, because being easy to set up means that others will probably already have done so, or will do if they see someone else making a success of the idea. The market size is also restricted to those with an interest in that form of research. For the R&D company the set-up costs are considerably greater and the research process will 'burn' through funds for months, or more likely even years, before a product is ready to be marketed. However,

Figure 7.3: Cashflow comparisons. ▶

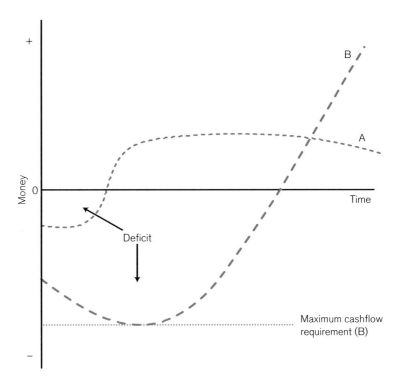

offering a patented product which meets a distinct market need means that company B is anticipating a rapidly growing and profitable market share.

Both businesses have funding deficits and will need to look outside for support. In the case of the newsletter business, the deficit could be covered by a number of sources (including a basic bank loan), but for the R&D-based company the options are limited by the scale of investment required and the duration of deficit involved. It is likely that 'staged' investment will be used, so that the investor does not necessarily have to meet the entire deficit in one payment but instead delivers funding dependent on successfully achieved milestones in the company's development. A well developed business plan can provide the investor with assurance that achieving the milestones will actually add up to a profitable business; this is essential for bio-businesses which may be in deficit for periods of 10 years or more.

7.8 – The entrepreneur's offer

For a business plan focused on attracting investment, this is the point when all the analysis and projections come together to form an attractive investment proposition. The plan's cashflow forecast will have revealed the extent of the funding deficit required and the financial forecasts will have demonstrated the future profitability of the business. All that remains is for the entrepreneur to ask for the level of funding they are seeking.

Investors might be offering loan or equity finance, possibly even a combination of both. The business plan must be tailored to meet the needs of these different sources of funding. For both debt and equity funding there are some common principles that will hold good for an attractive offer.

- *That the entrepreneur has made a clear personal commitment to the project.* Has the entrepreneur sufficient confidence and commitment to the business to sink their own money, as well as time and hard work (sometimes called 'sweat equity'), into it?
- *That risk is spread.* Although some equity investors may prefer not to involve others, it is generally perceived that a range of investors committing to a project is a good thing. In particular, lenders welcome sharing the burden of risk with others. If the entrepreneur can gain commitment from a reputable lender or investor for only part of the funding needed then this can help provide 'leverage' for funding from others.

For equity finance the situation is more complex than simply asking for a loan at a rate of interest that can be paid back within an agreed timeframe. The entrepreneur is essentially offering a stake in the ownership of the business, expressed in terms of a percentage number of 'shares' or equity. The incentive for the investor is that these shares will grow in value at a rate that far exceeds the many, much safer, forms of investment open to them. It is important to remember that an equity investment in this context is only worthwhile if it can be taken out of the business at some point in the future when it is worth considerably more; the so called 'exit route'. This means that the business plan must establish a strong case for what the company will be worth at the point of exit and how this exit will be achieved (see also *Chapter 8*).

For example, an entrepreneur anticipates that his business will be worth £20 million in a trade sale in five years' time. Throughout the business plan he will have provided market information to support this assumption and may even have given some examples from similar or related industries to illustrate the point. The entrepreneur believes a trade sale to a larger competitor is likely at this point. The business plan has identified a current funding deficit of £4 million. To raise this amount the entrepreneur offers an equity investor a 100% return on their investment in five years time, i.e. £8 million – this is equivalent to 40% of the business's potential value at that point. He could therefore ask for £4 million today in return for a 40% stake in the business. Naturally, the investor would argue that the business might well not be worth £20 million and so he should receive more than 40% for his investment!

Of course, people's perceptions of risk vary and market trends can be interpreted in all sorts of ways. Established investment industry expectations on timescales for realising investments are particularly problematic for the bioscience industry. Expectations tend to be around 3–5 years which, for many bioscience products, is too short a period in which to expect commercial results. Specialist investors in the sector, who are prepared to risk their investments over longer timescales, have emerged only recently.

The entrepreneur's offer may form the basis for future negotiations on the phasing and level of an investment or it might be rejected as unrealistic by the reader. Even if the investor accepts the logic of the offer and supports the investment in principle, there is then a process known as 'due diligence' that has to be worked through. Due diligence involves the detailed analysis of the entrepreneur's claims and evidence to ensure their validity. The background of the entrepreneur and the business team are also closely scrutinised to assure the investor that they too are credible. At this stage any claims made for the science and technology involved will be tested by objective experts. In short, this is a potentially lengthy and costly process that a good business plan can help mitigate.

7.9 – Managing risks

Investors accept some element of risk as a trade-off for a higher return than they could achieve by leaving their money in an interest-bearing deposit account. This is why good, experienced management is so highly prized by investors, as it will be the managers of the new business who will guide it through the inevitable problems it will face. A business plan should identify the main risks to the business and the likely strategies that will be used to deal with them should they occur. This will help anticipate investor questions before they can be asked and will help build the credibility of the plan.

Whilst developing the business plan it is likely that the author will have identified many risks to the narrative that they are establishing. It is inevitable that the investor will do the same and so it is important to order these risks in terms of their likelihood against the seriousness of their impact on the business. The investor will be looking for critical risks with a high likelihood of occurrence and it is these that the plan should address directly. Strategies for managing risks can consist of those for avoiding them all together or for mitigating their impact. A summary table, such as *Table 7.2*, may help with this.

Table 7.2: Summary of risks

Risk	Probability (P) 1 unlikely – 5 almost inevitable	Impact on business (I) 1 minor – 5 critical	Index (P+I)	Avoidance/mitigating strategy
Competitor product reaches market first	2	2	4	Increase R&D spend if evidence emerges this might happen If it does happen, the marketing strategy will be revised to highlight advantages of our technology over that of competitor
Year 1 US sales over 20% less than anticipated	3	4	7	Engage experienced US distributor with established contacts or withdraw from US market and refocus resources on EU
Costs of raw materials increase by over 5% during R&D phase	3	2	5	Source materials directly rather than through a distributor Revise introductory price of product in line with increased materials cost

Inevitably, the most frequently identified risks are concerned with lower than anticipated sales and/or margins on sales. Whilst it is possible to do quite precise homework on the technical and cost-based challenges the business will face, it is the concept of forecasting sales that will always be fraught with uncertainty. Providing a sensitivity analysis on sales will help address this issue, as will providing the reader with a break-even analysis

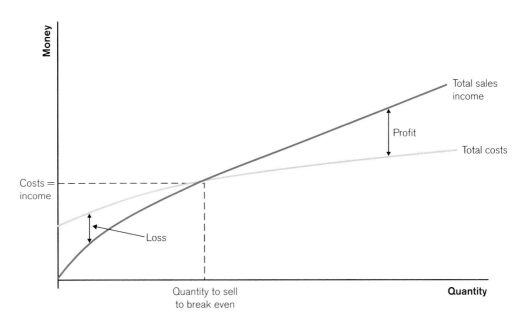

Figure 7.4: A break-even analysis. ▲

(see *Figure 7.4*) which shows that the entrepreneur is aware of exactly how many sales have to be made before a business can move into profit.

Risk management strategies depend on being able to see risks coming, or at the very least knowing when they are having an impact. As we saw with the development of an operational plan, building milestones into the business plan offers review points both the entrepreneur and the investor can use to determine whether any of the identified risks are affecting operations. For example, if a business is performing below forecast in a certain market then the planned review point offers the opportunity either to alter strategy or, in extreme cases, to close the business before any more investment is committed.

One risk that is peculiar to bio-business and other high technology sectors, is the dependence of the business on key individuals. Very often the involvement of the originator of the science behind the business is critical to the early stages of the business's development. Venture capitalists know this and may ask for 'co-sale rights'. This mechanism comes into effect if one of the business's founders attempts to sell their shares to a third party. With co-sale rights the investor has the right to replace a proportion of the shares held by the founder with their own, thereby giving them the contractual right to sell their shares along with those of the founder. The effect is to tie founders into the business. This is not usually popular with the founders.

7.10 – Concluding section

If there are vital messages that the plan needs to communicate to the reader then these should have been delivered in both the executive summary and the main body of the plan. However, there is no harm in making these points again in the conclusion. The concluding section should briefly draw together the elements of the plan, focusing on:

- the main reasons why the plan is likely to succeed
- the funding requirement
- the benefits to the investor of the offer presented
- the timescale in which action needs to be taken

The main aim is to encourage the reader to do something in response to the plan; hopefully to open more detailed negotiations on investment.

7.11 – Appendices

This final section should contain all the supporting information that is too lengthy and detailed to form part of the main body of the plan. While a business plan should be 'as long as it needs to be', in practice most, to spare the poor reader, do not exceed 40–50 pages. Appendices allow the inclusion of a great deal of valuable supplementary information that reinforces the plan without disrupting the flow of the main body of the text.

Information to be found in this section might include:

- full curriculum vitae of the management team
- specifications for the technology to be implemented
- patent documents and letters
- customer feedback letters or letters of support from other investors
- detailed market research reports
- supporting financial data

It is important for the overall standard of presentation that the appendices are put into the same format as the main body of the plan and are thoroughly checked for mistakes and inaccuracies. An investor finding an error in the supporting information would be justified in assuming that the conclusions drawn from that information may also be inaccurate.

7.12 – Confidentiality

Much of the information contained within a business plan is valuable both to direct competitors and to other businesses in the industry. Even the information that a start-up business plan exists might be enough to alert competitors and seriously disadvantage the enterprise's future prospects. Unfortunately, omitting sensitive information from the plan is unlikely to impress a potential investor who will want to see the whole picture. The author may therefore decide to restrict the number of copies of the plan in circulation and to ask recipients to sign a confidentiality agreement (see *Chapter 3*). Like patents, confidentiality agreements can only apply to information which is not 'in the public domain'. The entrepreneur should therefore choose advisors and other collaborators carefully.

7.13 — The finished business plan

Having laboured through the development of a full business plan, the author must now be prepared to alter it many times before it can be fully effective. For example, it is important to take account of valuable suggestions from trusted advisers. In addition, while some components remain the same, the plan must always be tailored to fit the specific interests of an intended reader. Remember that a business plan will very rarely get more than one chance to make a good impression on a potential investor. As Chris Matson at Medcom (*Case study 7.1*) recalls, the clear message from the feedback that he and his colleagues received on the first draft of their business plan was not at all positive: "Our early plans needed a lot of refining and we needed to listen to experienced advisors over quite a number of drafts before it was ready for investors to look at."

Once your business plan is complete, you will have a clear idea of the level of investment you need to start up your company. The following chapter describes the types of funding you should consider, and the ways you can use a business plan, and other devices, to characterise your business for potential investors.

7.14 — Additional resources

Finch B (2006) *How to Write a Business Plan,* 2nd edn. Kogan Page, London. Straightforward introduction to the form and content of a business plan.
Franklin C (2005) *Why Innovation Fails.* Spiro Press. A very readable look at the pitfalls of trying to get 'good ideas' successfully to the market. The sections on the dotcom failures are particularly relevant to business planning.
Haig M (2005) *Brand Failures.* Kogan Page, London. An entertaining review of business activities that failed to find a market.
Nokes S *et al.* (2003) *The Definitive Guide to Project Management.* Prentice Hall, Harlow. For those who don't know what PERT and Gantt charts are, this will tell you.
Sahlman W *et al.* (1999) *The Entrepreneurial Venture.* Harvard Business School Press, Harvard. This is an extremely comprehensive collection of readings on developing new businesses in the technology sector.
Stutely R (2002) *The Definitive Business Plan.* Prentice Hall, Harlow. Practical, step-by-step guide to putting a business plan together, illustrated with some fascinating examples.

www.bbsrc.ac.uk/business/

The Biotechnology and Biological Sciences Research Council website has a specialist business section with a range of case studies.

www.bizplanit.com/free/

Aimed at those preparing plans for venture capital.

www.businesslink.gov.uk

A guide to the general content of a plan.

7.15 – Acknowledgements

Thank you to the following for providing interviews: Mr Chris Matson (MD of Medcom), and Mr Nick Butler (Connect Yorkshire) for providing a venture capital perspective on business plans.

Learning outcomes

Key learning points from this chapter are:

- Most entrepreneurial business plans are written to attract investment. They should therefore be written to address the interests of the investor

- Investors are looking for a plan to present competent, experienced management and clear routes to profitable markets. The science involved is not, in itself, a priority – what the results of the science mean in the marketplace are

- Business plans rely on in-depth market research and analysis. Investors are looking for market conclusions to be supported by information referenced from objective, authoritative sources

- The financial projections presented by the business plan should be drawn from its market analysis. A sales forecast is usually the starting point for the other financial projections

- Accurately forecasting cashflow is essential to the survival of a business. It also provides an indication of what level of investment is required to enable the business to start up and grow

- An investor needs not only to be confident of a healthy return but also to know when and how this return will be realised. The business plan should make this 'offer' clear

- No business is risk free. The business plan should identify the major risks and show how they will be mitigated

- A business plan is a 'working' document. It should constantly change and be updated

Chapter **8**
Funding your Ideas

David Baynes and Louise Pymer

8.1 – Introduction

This chapter provides advice on how best to fund a business proposition. Sources of funding and the different sorts of people who may invest money are described. To increase the likelihood of them investing in your ideas, consideration is also given to the important things you must communicate to these people.

8.2 – Types of funding available and how to identify the right funds for your business

Before you start to approach potential investors, you need to know how much money your new business needs to prosper in its early years. Your business plan (see *Chapter 7*) will detail this amount, but it is worth remembering that you are not likely to achieve your business plan exactly and so it can be a mistake to go for too little money at the outset. Your business is likely to undergo several rounds of funding, but it is important that the first round gives you sufficient time to develop the business in a reasonable way, otherwise you will end up giving away too much equity during second stage funding.

You must also understand the nature of your business and how you intend it to generate money, both for you and any investors. For example, if your business plan shows only a small (<£100 000) cash requirement and then the generation of cash and profits, you may prefer to retain all the shares in the business and just borrow money for a short time – this leaves you with full control of the business and you will be the sole beneficiary if you sell the business at some stage in the future. It is more likely, however, that you will need more money than you can easily borrow and so you will need to sell some of the shares in your business to an investor and they will want to know your plan for an 'exit route' (see also *Chapters 6* and *7*); basically, when and how they will be getting their money back.

There are several types of potential investor that you may approach, but you need to be aware that they all have restrictions as to the amount of money they will want to invest. The following list can be used as a rough guide as to who you should be approaching based on the funding requirements identified by your business plan:

- friends and family – typically willing to invest no more than £10 000 each
- banks – with appropriate security they might lend you up to £100 000
- business angels – typically invest £10 000 to £250 000
- venture capitalists – typically only interested if they are investing more than £1 million

- institutional investors – not often an option for a start-up, but an option if, at some stage, you need several million pounds

Having identified who you need to approach, the following sections describe the pros and cons of each type of investor.

8.2.1 Friends and family

The great advantage of friends and family as investors is that you know them and they know you. The great disadvantage of friends and family as investors is that you know them and they know you.

The fact that friends and family know you, and hopefully like you, makes it easier to discuss investment with them and raise some cash. However, at the same time, your familiarity with these individuals may be a very good reason why you don't want to ask them for money in the first place. You may, after all, lose all of the money invested and this may well affect your relationship with these friends and family members. There are inevitably a number of complexities that can surface including the guilt you will feel if the investment fails or any form of dispute develops. There is also a risk that the whole transaction will be conducted on a fairly amateur basis without formal or legal documentation; this may lead to problems at a later stage, even if the business is successful. You will often have as many problems dividing spoils from successful investment, as you will have over money lost from a business. Investment by friends and family can often be problematic.

As a general rule, investment by those close to you must be discussed at length; whatever happens, it is likely to have a permanent effect on your relationship. It is very important, despite the trust that friends may have in you and your proposal, that you set up a formal agreement regarding how the money will be repaid, when, and how much. Clearly, however, if you have very good relationships with these individuals and you are borrowing only relatively small sums of money, say between £5000 and £10000, then this form of investment has the great advantage that it can be achieved without lots of complicated paper work. To borrow a relatively small amount such as this from anyone else would probably cost you a further £5000 in time, effort, interest, and legal advice.

If you are seeking investment at levels greater than £10000 then it would be best to explore more professional sources of funding.

8.2.2 The bank

The major benefits of approaching banks is that they are very accessible and have copious amounts of money. The disadvantage, as a general rule, is that they often won't want to lend it to you.

Banks do lend business people money, but only if the people they lend it to offer some form of security, usually in the form of a personal guarantee or a charge over their house or some other asset. If your business fails and cannot repay its loans to a bank, then the bank will come back to the individuals in the business who provided the security and ask them for their money back. Banks take security against loans because they do not like taking risks. Typically, banks will lend you money when you don't need it and won't when you do. In general, for the small borrower, the boot tends to be on the bank's foot, hence the adage: "If you owe the bank £10 000 and you can't repay it, you've got a big problem. If you owe the bank £100 million and can't repay it, the bank's got a big problem."

To be fair to banks, they are often very supportive of early stage businesses and will provide routine cash handling and deposit facilities. Their ability to provide overdraft facilities, so that businesses can survive cashflow problems for short periods, is important when debtors are slow to pay and creditors need to be paid. However, they must not be viewed as equity investors (equity investors often invest a lot of money by buying shares in a business rather than lending money that must be re-paid). Banks charge relatively small amounts for their money, often only 3–4% above base rate for a small, early stage borrower. This rate compares favourably with a venture capital investor (see *Section 8.3.4*) who will be looking for a return on their equity investment of around 25–45% per annum – venture capitalists look to double, triple or even quadruple the amount invested over a 4–5 year period. Banks can afford to take this much lower return because they rarely have to write-off much of the money they lend. Venture capitalists, on the other hand, assume that most of the investments they make will end up losing money. They must therefore ensure that the investments that do pay-off earn multiples of the original amount invested.

When you set up your business, you will need a bank to provide you with routine cash handling and deposit facilities. While setting up your account, it is always worth asking how they can help:

- they may lend you money to buy stock if orders have already been placed with your company and the bank knows you will make a profit out of the sale of that stock
- if you are willing to invest some of your own money then the bank may be willing to match that investment with an equivalent loan
- they may offer to work with you to apply for money from the Small Firms Loan Guarantee Scheme (a government-backed scheme where the bank takes most (70–85%) of its security from the government rather than you as an individual)

It is always worth 'shopping around' for the best deal regarding the terms of a bank loan and the facilities provided. The bankers will view this favourably as an indication of business acumen, as will the other investors you are looking to impress.

8.2.3 Business angels

Business angels may consider an investment of anything from £10000 to £250000. This, of course, makes them interesting potential sources of money for early stage companies. Business angels are often people who have been successful in their own businesses and now have the time and money to invest in other people's companies. They tend to be very confident and are likely to believe their advice is worth as much as their money, though often it is not. Always remember that it is their money you are interested in, although the advice you obtain will often be of at least some help to you. Recently, business angels have been supported by the new Enterprise Capital Fund money that the UK government has provided (see www.sbs.gov.uk/financegap/). This should mean that, over the next 4 or 5 years, there will be more 'angel money' available than has traditionally been the case.

An advantage of dealing with business angels is that they are more likely to look at a wide range of businesses and will be interested in anything that they think is a good idea and will make money; this will be particularly true if the business is in an area where they have experience and expertise. Business angels are therefore almost certainly the best port of call for many new businesses. They are often more helpful than venture capitalists because they have invested their own money. The drawback is that if you go to an angel, or a group of angels, it is likely they will end up as relatively significant shareholders in your business.

8.2.4 Venture capitalists

Venture capitalists have a bad reputation. It is not appropriate to say that all venture capitalists are bad, but it is fair to say that they have a relatively aggressive approach to investment and will expect a very significant return on the money they invest. Venture capitalists will be prepared to take risks but, because of this, will generally not invest unless they can see that there is real potential to make many times the original investment. A venture capitalist is therefore unlikely to invest in something that looks solid, but unspectacular. For example, if you were to set up a burger van on a popular street and you were able to sell enough burgers to make £400 per week, good for you. However, a venture capitalist will not be interested in investing in your business because although the chances of making a respectable return are good, there is not going to be a spectacular return on the investment. Conversely, you may produce a brand new medicinal compound that has the potential to cure AIDS but requires 15 years of work and an investment of hundreds of millions of pounds. The compound may have only a 1 in 100 chance of success in the pharmaceutical drugs market. Nonetheless, a venture capitalist may well be interested in investing a few million pounds in the early stages of this company, on the basis that if the compound is successful, then it has the potential to be worth billions of pounds and the venture capitalist would then make a very significant return on their money.

If your business needs initial funds of more than £1 million, and access to more money as the company develops, then venture capitalists are the people you need to speak to. You should be aware that your approach to them will be one of many they receive each day and as a consequence you should not expect a quick response. The following scenario is not uncommon:

- you produce your business plan, send it to your preferred venture capital group, and they sit on it for two months
- they may then find time to read it and suggest a meeting in a month's time
- you meet, they go away and think about it, and come back after a further month
- they finally send you a term sheet, which is a brief summary of what they are prepared to invest and their terms
- you argue about this for another month, then agree to a final version of the term sheet
- a legal agreement is drawn up (often at your cost) and you argue about the fine detail, via lawyers, for another month, but finally you are all set to sign up
- then you get a call from the venture capitalist the day before signing to say that he is only going to invest half the amount he promised - you argue but realise that you are going to go bust without his money, and grudgingly accept these new terms

So, it is generally not worth approaching venture capitalists unless you have something that has the potential to make very large returns. It is also worth approaching several at the outset so that you have other offers to fall back on.

8.2.5   Institutional and public market investors

The major institutional and public market investors invest their money in companies whose shares are traded on stock markets such as AIM (the Alternative Investment Market) or the London Stock Exchange (LSE). The advantages of trading your shares on these stock markets is that they often allow you to raise more money at a higher price per share (a better 'valuation' – this is the value of the company before the investment), from investors who are not really interested in interfering with the daily running of the business, than the other sources discussed here.

The disadvantage of AIM and similar markets is that you usually need to be a relatively well-established business before you can raise money by this route. Indeed, it is rare for a true start-up company to go directly to AIM to raise capital (though see *Case study 8.1* for an example). The reason for this is that these public markets expect to see a lower level of risk than other early stage investors such as venture capitalists. So the public markets normally expect to deal with a business that is established, hopefully with demonstrable turnover and already generating profit. An additional disadvantage is that an IPO (Initial Public Offering – the process by which your shares become listed and made available for anyone to purchase) is expensive; it can cost up to £1 million to raise £10 million.

Despite these costs, the stock market can be a good place to raise money.

- Although businesses in the public markets are exposed to detailed scrutiny, it may be easier to handle this scrutiny than that provided by the average venture capitalist. The venture capitalist is likely to get very tough with the management team if the business fails to deliver and it is not unheard of for founders to be removed from their own company within a year of taking money from a venture capitalist.
- The shares can be traded which reduces the risks for the investor. Effectively, if an investor decides that they have made a mistake, then they can sell their shares. However, if a venture capitalist decides that they have made a mistake with a business they have invested in, then there is no way for them to get out with their money. Worse still, they may find out that the company needs more money than expected and the only things they can do under these circumstances are invest more, or potentially lose their initial investment. Companies are therefore given better valuations on public markets because of the lower risks associated with this form of investment.
- Large amounts of institutional money are available from investors on the public market; for example, most pension funds are likely to invest in this way. This means that the public market is a very good place to raise large amounts of money. Another great advantage of the public market is that you can go back for more. If your business is floated on the public market, then you have a ready made mechanism by which to raise additional finance if you require it. This doesn't mean that the market will give you money if you are a bad company but, if you are prepared to set the share price at the right level, then you can go on to raise significant additional funds during what is called a 'secondary offering'.

As described earlier, AIM is generally not a suitable market for brand new companies although there are examples of true start-ups going straight to public market (see *Case study 8.1*). However, this is usually only possible if the start-up has some very valuable core IP or assets that have such remarkable potential that people are prepared to take additional risks. Investment of this nature occurs most frequently for biotechnology companies, although even in this sector public market investment in start-ups is rare.

Case study 8.1 – Syntopix
A start-up business that floated on AIM

Syntopix is a spin-out company from the University of Leeds, formed in 2003, and based on the research of husband and wife team Drs Jon Cove and Anne Eady. Their work focuses on the microbiology of skin and the development of safe and effective new topical treatments for conditions such as acne vulgaris and skin infections caused by *Staphylococcus aureus*. The problems caused by MRSA (methicillin-resistant strains of *S. aureus*) are well-documented and the market for the topical treatment and prophylaxis of *S. aureus* infection and carriage is around $0.5 billion. The global dermatology market as a whole is worth around $10 billion and there is an urgent need, within this marketplace, for new and more effective drugs to treat acne and antibiotic-resistant bacteria. Syntopix have developed a strategy that is designed to reduce the risks and costs associated with drug discovery by focusing on single and combined use of currently available compounds. The compounds selected already have a safe use in humans and this permits their fast-track progression through pre-clinical and clinical development. The strategy has inspired confidence in backers to the extent that only three years after start up Syntopix took the highly unusual step, for a new bioscience company, of floating on London's Alternative Investment Market (AIM) raising £4 million and achieving a market capitalisation of £10.1 million.

In recent years a new type of institutional funding model has appeared on the UK scene. This model is concerned with commercialising the IP that is generated at the country's leading universities. A number of investment companies have obtained funding from the public markets which they secure in exchange for the rights over the IP at partner universities for some predetermined period going forward. In effect, what they are doing is raising money initially by selling the future stream of IP being generated at the partner universities, and then using that money to actually commercialise those ideas as they appear. These companies combine the activities of technology transfer, company incubation and early stage venture capital funding.

The leader in the marketplace is a company known as IP Group (formally IP2IPO) who now has arrangements with 10 universities in the UK. Under these agreements, IP Group sets aside a ring-fenced fund normally valued at about £5 million, in exchange for rights to approximately 12% of all new start up companies for a pre-agreed period, often 25 years. IP

Group are one of the funders of Syntopix Group described in *Case study 8.1*.

The other major players in this market place are:

- BioFusion – has agreements with Sheffield University and Cardiff University under which it owns all of the IP generated at these universities; BioFusion has a right to own all shares except those given to the founding academics in start-up companies that are formed over a period of 10 years. This model is very attractive to universities, which are able to participate in the success of the business, and it seems likely that more universities will join the Biofusion model over the coming years
- Imperial Innovations – is responsible for commercialising all the IP being generated at Imperial College in London

Each of these 'technology transfer' companies is at an early stage of development. However, they have already shown some very impressive results with nearly 100 companies being started-up and sold or floated in the last five years. Imperial Innovations alone has floated one business worth more than £150 million.

This is an exciting new model and one which organisations in the rest of the world appear to be keen to mimic. The limitation of course is that they only provide funding for the universities with which they are associated. So if you are not operating out of a university with which these organisations have a relationship you will not be able to access the funding in question. If, however, you are lucky enough to be situated at one of these universities, then your local technology transfer department will be able to point you in the direction of the companies concerned. If you can present a sensible business plan backed up by properly secured IP you have a very good chance of raising funding. Importantly, you will also be able to access the expertise that these companies have in funding, building, floating and selling businesses.

8.2.6 Grants

Grants can be excellent for business by providing what is often referred to as 'soft money' – money that you do not have to pay back. A major disadvantage of grants as a source of funding is that grant application processes are often tortuous, requiring a large amount of time on administration, and often a certain amount of skill in knowing exactly how and where to target the proposal. Of course, at the end of this lengthy procedure, there is no guarantee that the application will be successful. However, free money is a good option for a business and so, where grants are available, you should look to submit an application. No successful start-up can be dependent on such funds. A successful business should be able to survive without them.

You should be aware that as well as the more obvious sources of funding for the biosciences, like the Medical Research Council, Biotechnology and Biological Sciences Research Council, and the Wellcome Trust, there are many smaller organisations that may fund your ideas. You may also be able to obtain funding through Regional Development Agencies and Knowledge Transfer Organisations (see *Sections 8.4.4 and 8.4.7*).

8.2.7. Competitions

To encourage innovation and investment a number of public bodies, with government funds, run business competitions. These can be useful in providing risk-free financial support if you are successful, though the sums involved are relatively modest. However, they can generate important 'seed-corn' funds for more risky ventures at an early stage. Probably the best reason for entering competitions during the early phases of developing a new idea into a business is that they provide a means of getting you noticed and kudos if you win. Even if you are not successful in the competition, potential investors are likely to be involved in choosing winners or generally monitoring what the competitors are proposing. However, the publicity and exposure in competitions has a downside – your competitors will become aware of your enterprise and you will have to protect your IPR situation (see *Chapter 3*).

8.3 – Selection criteria used by different sources of financial backing

As a general rule most potential investors use the same criteria for evaluation of an investment opportunity. The main criterion is: will I make money from this investment? This is closely followed by issues like: can I trust the people I am investing with?, do I like them?, and am I excited by the technology?

This section briefly considers the different criteria that will be used by each of the main sources of capital investment that we have identified.

8.3.1 Friends and family

These people already know the person they are considering investing in. They are likely to consider factors that differ from the criteria used by other groups of investors. The questions they will ask include:

- will it be awkward to say no?
- am I prepared to invest and, perhaps, lose money whatever the suggestion because, for example, my son or daughter is asking me to invest?
- do I like the person who is asking me to invest?
- will I make money out of the investment?

8.3.2 The bank

As already discussed, banks do not make equity investments and so their selection criteria will be based on the following:

- is the bank completely and utterly convinced that the loan will be repaid?
- is the borrower providing sufficient security so that the bank is covered even if all of the money is not repaid?
- can the bank charge adequate interest on the loan?
- will the bank have the opportunity to do transaction work for the customer?
- can the bank trust the borrower?

8.3.3 Business angels

Business angels tend to be obsessive in their selection process and their criteria are likely to be based on the following:

- do I have specialist knowledge in this area and so can I add value to the proposition?
- do I think I've identified something here that no one else knows about?
- do I like and trust the individual asking for support?
- am I excited about the technology?
- am I going to make money?

8.3.4 Venture capitalists

Venture capitalists are likely to be more professional and even more hard-nosed than business angels. The kind of selection criteria they will apply include:

- is this an industry or sector in which I normally invest?
- is the amount of money requested at a level of investment I am comfortable with? (a venture capitalist might reject a proposal if the amount requested is too little or too large)
- do I trust the individuals asking for support?
- do the individuals have significant experience in this area?
- am I excited by the technology?
- am I going to make a remarkable return on this investment?

8.3.5 Institutional and public market investors

Some of the criteria applied will differ from those used by the other groups of investors:

- do we believe that this company will grow and generate profit and provide us with dividends?

- is this a company that we understand and feel comfortable having in our portfolio?
- do we trust the management team?
- is this a sector in which we would normally invest?
- are we already overexposed in this sector (i.e. are we already heavily reliant upon investments in this field)?

In summary, it should be clear that some of the criteria listed above are applied by all of the groups of potential investors. For example, for most investors it is essential that the proposition they are being asked to support has a very good chance of making significant amounts of money. In addition, you should now be aware of all of the further, more specific criteria that will be used by banks, business angels, venture capitalists and public market investors when you approach these organisations for investment and support.

8.4 – Sources of advice on funding and opportunities to engage with potential backers

There are a number of sources of advice on finding funding and various opportunities to meet potential backers. These sources of advice and meetings will be of varying degrees of usefulness.

8.4.1 Talking to a bank manager

You will work closely with your bank manager even if only with regard to day-to-day transactions, and so it is worth talking to a few to ensure you find someone you are comfortable working with. While your bank manager will not make an equity investment in a technology company, they might be able to direct you to someone who is prepared to invest.

8.4.2 Local angel groups

Most regions of the UK have 'angel' groups, including:

- Advantage Business Angels (Midlands and national – www.advantage businessangels.com)
- Archangel Informal Investment (Scotland – www.archangelsonline.com)
- Business Investors Group (North East – www.big-angels.co.uk)
- Cambridge Angels (www.cambridgeangels.net)
- Envestors (London and South East – www.envestors.co.uk)
- London Business Angels (London, home counties, Oxbridge – www.lbangels.co.uk)
- SWAIN (South West – www.swain.org.uk)
- Thames Valley Investment Network (www.tvin.co.uk)
- Xenos (Wales – www.xenos.co.uk)

- TechInvest (North West – www.techinvest.org.uk)
- Yorkshire Association of Business Angels (includes Humber region – www.yaba.org.uk)

8.4.3   Venture capitalists

Venture capitalists often seem somewhat elusive; they are, in fact, very easy to identify – the British Venture Capitalist handbook is available online (www.bvca.co.uk) and includes full details of all the major venture capitalist houses in Britain. This publication also provides extremely useful information about the sectors that individual venture capitalists prefer to invest in and gives an indication of the minimum and maximum investment levels for each. Further information is provided about other investments the venture capitalists have made. This handbook should be used as a starting point for the distribution of your business plan.

8.4.4   Regional development bodies

There are a number of regional development bodies which manage European funds worth many millions of Euros. Much of this money is made available in the form of grants. Make a point of contacting your own local regional development body to determine how grant funding can be obtained in your region (see, for example, www.yorkshire-forward.com, the regional development body for Yorkshire).

8.4.5   Professional grant advisors

These firms provide individuals and companies with professional advice on how to obtain grants. The advantage of taking this approach is that these companies are aware of all of the relevant grants available, and will provide invaluable advice about how to compile an application. The downside is that they can take up to 25% of the money received.

8.4.6   Public funding

The public market can only be accessed through professional brokers and corporate finance advisors. These advisors range from straightforward accounting firms, to boutique advisors (small independent corporate finance advisors), to the large American investment banks. For a list of approved advisors, contact AIM (www.londonstockexchange.com/aim) or the London Stock Exchange (www.londonstockexchange.com).

8.4.7   University technology and knowledge transfer offices

Most UK universities have well-established research support offices that will provide valuable advice about sources of funding. Your university will be able to advise you on a wide range of sources of grant income from research councils, charities, and regional development agencies. As discussed in *Section 8.2.5*, some will already have agreements in place with investment companies that can provide a great deal of useful information and, potentially, early-stage funding for bio-businesses.

Now that you are aware of the different types of funding available you can decide on the source and level of support most appropriate for your

Simon Best
Ardana plc

Simon Best's career to date has been both exciting and inspirational but rarely predictable! Simon graduated with a BA in Music from University of York in 1977. His first job involved musical talent-spotting in Yorkshire and he was involved in signing The Human League with the up and coming Virgin label. However, in the early 1980s Simon decided on a major career change that was influenced by two important developments. First, while working in the music industry he discovered a talent for managing highly creative people. Secondly, a number of friends who were scientists made Simon aware of exciting developments in the emerging science of molecular biology, and the great potential for biotechnology worldwide. Simon therefore embarked on an MBA at London Business School that led to a highly impressive career in management and bioscience:

- During MBA studies, Simon and a scientist friend won a business plan competition based on the proposed isolation of active compounds from traditional indigenous and oriental medicines, and the development of patentable synthetic homologues.
- In the mid 1990s, at ICI, Simon managed the development and marketing of the first GM product (tomato purée) sold in UK supermarkets.
- In 1998, Simon was head-hunted as CEO of Roslin Biomed, the company formed to

exploit medical applications of the cloning technology developed to create Dolly the Sheep; Simon rapidly developed the business and facilitated its acquisition by the Geron Corporation who wished to exploit stem cell-based technology for the treatment of the major degenerative diseases of ageing.

- In 2000, Simon was asked by the MRC Human Reproductive Sciences Unit (HRSU) to assess the commercial potential of their research.

Simon's involvement with HRSU led him to identify the massive potential for growth in the field of human reproductive therapy and he managed to obtain seed funding for Ardana Bioscience. As a private company, Ardana raised over £43 million of venture capital in three funding rounds between 2000 and 2005. It has a number of exciting drugs in the pipeline including Teverelix LA for prostatic diseases in men and endometriosis in women, and Terbutaline for endometriosis-related infertility. In March 2005, Simon raised over a further £20 million by floating Ardana plc on the main board of the London Stock Exchange. Simon is clearly a man of many talents and he draws interesting comparisons between his life as a musician and career in management. In each, he says, there is a need to be able to generate and track complex patterns and long-range structures, focus on fine details and yet keep in mind the overall score and composition.

business venture. You will then need to work on your presentational skills to ensure that your ideas are presented most favourably to potential backers.

8.5 – Characterising your business for potential investors

This section describes the variety of ways in which you can characterize your business using terms and presentation methods your potential investors will be comfortable with.

8.5.1 Mission statements

Mission statements for a start up company are widely considered to be inappropriate. Traditionally a mission statement is an attempt to summarize a business's aims and objectives in a single sentence or paragraph. It is believed that a mission statement provides focus for a business, so that all the company's staff can gain a clear idea of what the company's aims and objectives are. For this reason companies spend many months, if not years, arguing over the correct wording that ought to be used. The majority are unsuccessful.

In some circumstances a mission statement can benefit a company, particularly if the business is extremely complex and the statement is kept suitably simple. For example, a large international corporation might find it useful to communicate an idea of where it wants to go, and how it wants to get there, to its 150000 members of staff. This could be achieved by distributing a simple mission statement to all staff and displaying it in the reception area of all 300 offices around the world. It is in this environment that mission statements can be most constructive.

However, this book is not about companies with 150000 employees, but is about high-tech, early stage start-ups. Unfortunately, a simple mission statement trying to express the aims and purpose of an early stage start-up will usually fail to achieve its aims, and so it is unlikely to be of interest to a potential investor who needs to actually understand something about a company when it is first presented to him or her.

8.5.2 The elevator pitch

In an elevator pitch you have the equivalent of a one minute elevator/lift journey to communicate your idea to a potential investor.

Crucially, when selling an idea for a business during an elevator pitch, you must make plain how the business will make money and why the potential investor should be excited by the opportunity you are presenting. The concept of the elevator pitch is covered in more detail in *Section 5.3.2*.

8.5.3 The business plan

The business plan has already been described in detail in *Chapter 7* and so will not be considered at length here. However, it is important to stress that it is unreasonable to expect an investor to put money into your company if you cannot effectively communicate, in a business plan, what your company does and, most importantly, how an investor is going to make money from it. Business plans usually take rather a long time to finalise and that is often because companies, in their early days, are invariably a jumble of different ideas. Producing a business plan can therefore be a valuable exercise that will allow you to crystallise your thinking and help direct your plans.

Business plans always require an executive summary at the beginning. This is a summary of the main body of the plan and is short enough for the busy executive to read. Business plans can be as long as 100 pages and so it is unlikely that anyone will actually sit down and read the plan in its entirety. The aims and skills involved when creating a good executive summary are very closely related to those involved when creating a good elevator pitch. Some elements of the executive summary are best written before the rest of the document and some afterwards (see *Section 7.5.1*)

The structure for the business plan is described in detail in *Chapter 7*, but it is important to stress here that you should certainly include the following: what the business does; what IP the company owns now, or will own once it has completed the development process; what is unique about the business; the financial projections for the business, including its cash requirements; and details of the management team – this is perhaps most important of all (see *Section 6.2* for further details).

It is worth a quick mention of management at this stage. When they first receive a business plan, many venture capitalists turn straight to the management section. If the management team has experience, or is known and respected, the venture capitalist will read the plan. If the management team is completely green and the people have no experience, the plan may never be read. Good management is vital if you are to obtain funding for your idea.

8.5.4 The formal presentation

The elevator pitch may get you in the door, and the business plan may hook the potential investor's interest, but very few will invest until they have actually met you and heard your story. The investor is also unlikely to have read your business plan in any detail and you must therefore be able to describe your business opportunity through other means. This usually involves a formal presentation, often using PowerPoint technology. Some people are great presenters, but even they can fail if they prepare a bad PowerPoint presentation. It is therefore essential that you, possibly an inexperienced speaker, produce a good presentation that really summarises your business. A well-prepared presentation will help guide you through the key points and ensure that you stay with the subject. It should ensure that you don't leave out any important aspects. Without the discipline of a well-structured presentation, many presenters can spend a whole hour talking at length about, for example, IP, and forget all the other components that are important to potential investors.

You should appreciate that a potential investor may already have sat through a number of other presentations that day and so you need to make sure your presentation is visually interesting. Furthermore, a professional presentation projects you as part of a substantial professional organization, even if the reality is that you are actually an early start-up with only one employee.

Ensure that your presentation:

- is interesting
- is professional in appearance
- is not overburdened with text

But above all, your presentation must include all the key points about your business that you wish to communicate:

- that there is a real problem to be solved using your product or service
- that your business proposition will solve it better than the competition
- why your team is best placed to solve it
- who is going to buy the product or service
- how much money is going to be made
- how much money is required and how it will be repaid

For further, valuable advice about successful presentations see *Chapter 5.*

The audience for the presentation

You will often find that there is an inverse relationship between the amount of time you will spend raising money, and the amount you are trying to raise. This is counter-intuitive, as it would appear sensible to assume that if you are raising a small amount of money then you will spend less time on this than the amount of time you will spend raising a large amount of money. However, it is almost always the case that smaller sums of money take a considerably longer time to raise than larger sums, with this relationship holding true all the way up the money raising 'food chain'.

Business angels, despite priding themselves on their business acumen, will usually spend a long time in discussions with a company before they decide to invest their own money. They will often insist on numerous meetings and a large number of conditions being agreed to before they are prepared to see their money being invested. They may, for example, insist that you change your business plan, or ask that you try to license your product rather than sell it, and so forth. This results from the fact that most angels are successful business people and so know an awful lot more than a new, budding entrepreneur about what will create a successful business. In the end it will probably be 9–12 months before you get any money from them, but you could obtain up to £250 000.

Venture capitalists will also take a long time to persuade to fund your idea, but they will be quicker than a business angel. Once they have found the time to read your business plan and then meet with you, they will make a decision fairly quickly and so you could get your money after only 6–9 months, and it might be in excess of £1 million.

Floating your company on AIM or another public market is often considered by inexperienced entrepreneurs to be incredibly difficult. In many respects, however, the procedure is simpler and more straightforward than dealing with business angels or venture capitalists. Professional advisors should be employed to guide you through the process and they can organise up to 40 meetings with different potential investors over the space of, say, 10 days; during each of these meetings you will make a presentation lasting around 45 minutes. In each case, this will be the only time the investor will meet you before they make an investment decision. Those investors who decide to invest will put in, on average, between £0.5 and £1 million. So, for a 45 minute presentation you receive up to £1 million.

The reminders above are provided to emphasize the importance of the formal presentation. For all of these investors you end up giving a presentation. The business angel will expect a presentation, venture capitalists dine out on presentations, and you only get money from AIM and institutions by giving a single, high pressure presentation. Even for friends and family you will have to give a presentation, albeit much more informally, but you are still in sales mode, presenting and telling the person certain key things.

All formal presentations must address the following questions which are vital to investors (you will note that the answers to these questions will be very similar to the information that must be included in an elevator pitch or the executive summary of the business plan; see *Sections 8.5.2* and *8.5.3*):

1. what does your business do?
2. what unique IP does your company have and what particularly strong position does it have against competitors?
3. what is the market opportunity that the investor should be excited about?

4. who are your management team?
5. what are your financial projections and when and why is the business going to make lots of money?
6. what are you going to spend the investor's money on?

8.6 – Solutions to common funding problems

Problem. The parties cannot agree on the pre-money value (that is, the value before money is invested) of an investment.

Solution. There is no simple answer to this one as both sides will naturally want to negotiate the best deal they can. To improve your chances of obtaining the valuation you want, ensure that you identify more than one person who is prepared to invest in you, so that neither party can hold you to ransom.

Problem. No one wants to invest in you.

Solution. Keep trying. If you believe that you have a good proposition then persevere. Individuals rarely raise money easily, no matter how good their proposition. However, it is a good idea to make sure you listen to the advice that people give you. If five people indicate, one after another, that they will not invest because they don't like the colour of your product, change the colour.

Problem. You don't know how to produce a financial plan.

Solution. Make sure you have access to a good accountant. If the numbers don't add up, or don't make sense, then you will lose credibility at the outset. Investors will be very uncomfortable about handing cash to someone who doesn't seem to know how to keep count.

Problem. It is taking for ever to get terms agreed and the deal completed.

Solution. Find an alternative income whilst you work on setting up your new business; the process will probably take at least twice as long as you expect. If you have an alternative source of income it will keep the pressure off and will help ensure that you do not have to agree to a bad deal just to keep the wolf from the door.

Having worked hard to generate, protect, research, and now fund your idea and start your business, you face another, major, hurdle. The bioscience industry is very heavily regulated and the next chapter introduces the many regulations that you must understand and adhere to, if your company is to function as a commercially viable concern in the marketplace.

8.7 – Additional resources

www.bvca.co.uk
: The British Venture Capitalist handbook available on the web.

www.londonstockexchange.com/aim
: AIM on the London Stock Exchange site.

www.sbs.gov.uk/financegap
: Enterprise Capital Fund money that the government has provided.

www.yorkshire-forward.com
: Local regional development body for Yorkshire.

Regional business angel groups:

www.advantagebusinessangels.com
: Advantage Business Angels (Midlands and national).

www.archangelsonline.com
: Archangel Informal Investment (Scotland).

www.big-angels.co.uk
: Businesss Investors Group (North East).

www.cambridgeangels.net
: Cambridge Angels.

www.envestors.co.uk
: Envestors (London and South East).

www.lbangels.co.uk
: London Business Angels (London, home counties, Oxbridge).

www.swain.org.uk
: SWAIN (South West).

www.tvin.co.uk
: Thames Valley Investment Network.

www.xenos.co.uk
: Xenos (Wales).

www.techinvest.org.uk
: TechInvest (North West).

www.yaba.org.uk
: Yorkshire Association of Business Angels (includes Humber region).

Learning outcomes

Key learning points from this chapter are:

- The wide range of potential sources of funding available and how to identify the right source of funding for your ideas and business

- The selection criteria used by potential investors

- The need to always have an 'elevator pitch' (a one minute summary of your business idea) ready for use

- The importance of a coherent business plan with financials that add up

- The value of a high quality formal presentation that can be circulated in paper or electronic format alongside your business plan

- Don't give up!

Chapter 9
Regulation in the Biosciences

Nick Medcalf & Bob Pietrowski

9.1 – Introduction

9.1.1   What is regulation for?

Bioscience has immense potential to improve the quality of human life, but such potential is open to misuse. Companies that operate in a free market will be under competitive pressures, and reckless companies may be tempted to cut too many costs in an effort to compete. The result can be poor quality or dangerous goods and these must be prevented from entering the market place.

Governments impose regulations to protect consumers from this risk. Effective commercial competition takes place when all organisations are operating to common and appropriate standards. The regulatory authorities achieve this state of affairs when they create, police and maintain the standards within which the goods and services may be traded.

Regulatory controls are, naturally, most extensive and restrictive where the potential for harm is greatest. In the biosciences, these areas are:

- the creation, use, containment and distribution of genetically modified organisms
- handling and disposal of the waste products of bioprocesses
- goods coming into contact with the environment, e.g. pesticides
- animal science and welfare
- biomedical products

Regulations go beyond safety matters alone, with ethical (see *Chapter 10*) and business issues also being important. Government regulation takes account of these issues through representations made by interested parties. Consumer groups, professional bodies and government-appointed working groups (involving appropriate specialists) will make major contributions. Some of these groups provide excellent reference resources when they publish information about forthcoming regulations (*Table 9.1*).

Table 9.1: Some organisations offering information about bioscience regulation

Organisation	Country of origin	Website address
BioIndustry Association (BIA)	United Kingdom	www.bioindustry.org
Biosciences Federation (BSF)	United Kingdom	www.bsf.ac.uk
Biotechnology Industry Organisation (BIO)	USA	www.bio.org
Friends of the Earth (FOE)	United Kingdom	www.foe.co.uk
Japan External Trade Organisation (JETRO)	Japan	www.jetro.go.jp
Soil Association	United Kingdom	www.soilassociation.org

9.1.2   The importance of the regulations

The first encounter with regulations can come as a shock for the scientific entrepreneur. The public operating environment is very different from the research environment. The operator is accountable to governmental bodies and it is no longer enough to show that something is true scientifically – it must also be demonstrated and recorded. The origins of any assertions about the product must be traceable. In addition, the entrepreneur has to take the time to keep up to date with the regulations. The businessperson must also be creative, as situations sometimes arise where there may be no clearly defined way of meeting the regulations, and a balance must be struck between public and commercial interests. This is especially true for new technologies because they create situations for which no precedent may exist. An example of this occurred during the early 1990s with the advent of genetically modified foodstuffs. Regulations existed in the EU to cover the food use, but for a while the regulatory pathway for applications for animal feed or for seed production were less clear; ultimately each innovation was scrutinised individually. In this situation, the competitive edge will belong to the organisation best able to satisfy the regulators in a commercially sensible manner.

The business and the regulator both want the consumer to be provided with quality goods or services. The quality of bioscience products differs from that of traditional products in two ways: first, bioscience products are usually much more complex, and secondly, the methods of manufacture are often more indirect. Where living organisms are used in the process the final product may also have features that are not under the direct control of the operator.

Successful businesses need robust, controllable processes. Small changes occur even under normal, well-managed production. The ideal process will tolerate these and will not yield product that lies outside its specification. The process itself can then be treated as the dominant factor governing the product properties. To get a better idea of what this means consider an everyday activity such as baking a cake. Some people just seem to have a 'knack' for getting a sponge to rise every time they bake one, while others keep getting poor results. Why is this? Different ovens, ingredients, starting temperatures and so on lead to too many features that can go wrong. One way of getting a good result every time is to pin down as many of these features as possible and then, when a good result has been obtained, to find out which ones are important. The same, optimised, procedure can then be used every time: same oven, same brand of ingredients, same procedure. This is the basic approach taken in process control and comes under the general heading of 'process validation'. In process validation the development scientists must examine the process to find out which features exert the most influence over the product. They must then define the allowable variation in these features and show that the process does not stray from those in normal operation.

The regulatory authorities expect a business to know their process inside-out. New products usually have to be registered and the method of production will need to be registered as well. Any deviation from the registered process must be avoided. If any changes become necessary, the business must undertake studies to justify the change and to show that all the critical properties of the product remain unaffected.

9.1.3 The regulations

Regulations differ from country to country. Each country in the developed world has its own 'competent authorities' (the organisations responsible for enforcing compliance). In some cases there are agreements between nations to allow trade using common standards.

Within the European Community (EC) directives and regulations apply across all the member states. For readers planning on working in the UK it is good to be able to read the full text of these and they can be found at the website of the European Commission (ec.europa.eu).

In the USA, the Federal Government is responsible for the creation and issue of Federal Regulations. These are published as the *Code of Federal Regulations* (www.gpoaccess.gov/cfr/index.html), a central record arranged under 50 'titles', each of which covers a different subject. It is often necessary for non-US companies to comply with these rules if a product is to be imported into the US.

9.2 – The impact on the entrepreneur

9.2.1 Quality management systems

A company working in a regulated field needs a quality management system (QMS). A QMS is a way of working that ensures that the separate procedures within an organisation join up in an overall system to provide a consistent quality of goods or services. Anyone who has ever had work carried out on their house or who has tried to organise an event like a wedding will appreciate the large number of opportunities, even in such a modest project, for misunderstandings, mistakes and mix-ups. The same is true in bioenterprise where the development of a business takes place over years and involves lots of people, some of whom will move on to other jobs. The purpose of a QMS is to keep everything transparent and traceable. For a QMS to be effective it must join up all the resources going *into* an operation (people, training, purchases, etc.) with all aspects of the product coming *out*.

If a business plans ahead there will be every chance that the goods will be right first time. The underlying message of all QMSs is to keep to a controlled cycle of 'Plan – Execute – Record – Improve'. Sometimes mistakes will be made and corrections will be needed but this should not be used as a substitute for planning ahead. It is more cost-effective to follow this pattern than to try and make adjustments when orders for plant, buildings and materials have already been placed.

9.2.2 Costs and compliance

A new manager may be tempted to save time by skipping record-keeping, but this is nearly always a false economy. Regulatory requirements can stretch all the way *back* into the way goods were purchased and all the way *forward* into how the manufacturing will be carried out. (For biological products it is especially important to conduct trials to show that the transport arrangements for the goods will be sure to deliver the product in the desired condition.) To take one example: machinery with on-board computer controls is subject to standards that are published as 'Good Automated Manufacturing Practice'. As the equipment is designed, tested, approved for purchase and delivered to the site it should be subjected to tests that show it is fit for purpose, and records must be kept for later inspection. This approach requires more time and money than to 'just do it', but the alternative – to return and repeat the work at some future point – is usually much more expensive and sometimes impossible.

9.3 – An introduction to the regulations subject-by-subject

9.3.1 Genetically modified organisms

Genetic modification of organisms has brought the promise of better access to high value, low cost materials than would otherwise be possible. However, the creation and containment of genetically modified organisms (GMOs) must be done responsibly in order to protect public health and the environment. The authorities have been particularly vigilant over this topic.

In the UK the SACGM (Scientific Advisory Committee on Genetic Modification) is a key source of information for anyone working with GMOs. The Committee produces a 'Compendium of Guidance' (www.hse.gov.uk/biosafety/gmo/acgm/acgmcomp/index.htm) which is valuable when learning about what the regulations mean in the workplace. The key elements of the law are 'The Genetically Modified Organisms (Contained Use) Regulations 2000' and its later amendments. In the UK the Health and Safety Executive (HSE) has responsibility for the 'contained use' regulations.

In the UK, a GMO is any organism in which either the DNA or RNA of the host has been altered by man. Thus animals, plants and (most commonly) micro-organisms (GMMs) fall under the law. No establishment may operate in the UK using GMOs until it is registered. There are fees for this and registration is not permanent but must be renewed when activities change or develop (further details can be found at www.hse.gov.uk/biosafety/gmo/law.htm).

The risks posed by the GMO and the processing must be considered with regard to impact upon humans and upon the environment – GMMs in particular have the potential to act as disease agents unless features are built in that limit their ability to survive outside the bioreactor. Once the activities have been assessed, compulsory containment measures are then specified. The law is quite prescriptive about the need for rapid notification in the event of accidental releases.

Movement of GMOs between countries poses a special risk of disruption of ecosystems. The 'Cartagena Protocol on Biosafety' (see www.cbdint/biosafety/default.shtml) gives member states the right to advance warning of any importation of GMOs into their territory and the right also to make a decision about it in advance. More than 100 countries have so far signed the protocol.

The most useful guidelines in the USA are those contained in the Federal Register[1] and those provided by the National Institute of Health. These lay down procedures to follow and the best practice for their execution.

Among larger organisms, the category with the most impact to date is GM crops. In the USA, the Department of Agriculture (USDA) regulates this activity and the Animal and Plant Health Inspection Service (APHIS) inspects and controls field-testing of biotechnology crops. In the UK, the Department for Environment, Food and Rural Affairs (DEFRA) oversees this activity and is responsible for inspecting applications that involve introduction of genetically modified crops in field trials. During 2006, six such trials were conducted within Europe including one for a blight-resistant potato variety.

9.3.2 Genetically modified food products

Genetic modification of food, often carried out to improve yield, is controversial. There are differences in approach between Europe and America. The EU has issued two regulations:

- the first (the 'GM Food and Feed Regulation EC1829/2003') specifies how the marketing of such foods must be conducted
- the second ('Traceability and Labelling Regulation EC1830/2003') gives instructions about how the food labelling must be done

The duties extend into any subsequent foodstuffs that may be made from the original foodstuff. The intention is to create a data trail allowing any ingredient of GM origin to be identified by the customer.

Within the EU an organisation that creates a new foodstuff of GM origin must apply to one of the EU Member States for marketing preauthorisation. The Member State will then make the application available to the European Food Safety Authority (EFSA). The European Commission will prepare a draft decision based on the opinion of EFSA. A separate committee known as the 'Standing Committee on the Food Chain and Animal Health' (comprising experts from the Member States) will either confirm this draft decision or subject the application to further consideration. Any approval will be valid for a period of 10 years.

The position in the USA is somewhat different. New food products from biotechnology are regulated using the same mechanisms as those used for traditional foods under the 'Food, Drug and Cosmetic Act'. The primary concern is the *safety of the product* rather than the *method of production*. The responsibility for monitoring the introduction of GM foods is shared by several regulatory bodies: the Food and Drug Administration (FDA), the US Department of Agriculture (USDA), and the Environmental Protection Agency (EPA).

9.3.3 Pesticide activity through genetic modification

The other main reason for the genetic modification of crops is to enhance their resistance to pests. This feature may upset the balance within an ecosystem and therefore new developments are subject to regulatory scrutiny. In the USA, the legislative basis for this lies in the 'Federal Insecticide, Fungicide and Rodenticide Act' (FIFRA) and the 'Toxic Substances Control Act' (TSCA). These two Acts enable the EPA to control the invention, testing and distribution of any crops able to express pesticides, that are produced by biotechnology.

In Europe these products are addressed by the specific pesticide directives and those governing introduction of GMOs. Member States enforce the directives through whichever agricultural agency has a mandate to scrutinise pesticide safety, e.g. the Pesticides Safety Directorate in the UK.

9.3.4 Stem cells

Regulation of the therapeutic use of stem cells is a hotly debated subject. There are differences even within Europe and also within the USA with regard to the available public funding. The issue of the use of *embryonic* stem cells has usually been at the heart of the debate. The use of *adult* stem cells, i.e. cells that are derived from mature tissue but that retain pluripotency, has not attracted the same degree of controversy. However, many members of the public confuse the use of adult stem cells with the use of fertilised embryos for similar purposes.

In the UK the creation of embryos for limited research was made legal under the 'Human Fertilisation and Embryology Act 1990'. Later the 'Human Fertilisation and Embryology (Research Purposes) Regulations 2001' allowed use of such embryos for research where this research developed understanding of serious noncongenital diseases. The activity is tightly controlled by the Human Fertilisation and Embryology Authority (HFEA). Each project must be separately licensed and failure to obtain permission before conducting such work is a criminal offence. Cloning from fertilised eggs remains prohibited.

A ruling was made in 2001 that the technique of cell nuclear replacement did not qualify as 'fertilisation'. Such products do not therefore qualify as embryos and do not come under the HFEA remit. However, the 'Human Reproductive Cloning Act' of 2001 makes it an offence to attempt to place such an 'embryo' in a woman's body.

Within Europe local regulations vary widely. The EU does not currently regulate the subject of stem cell research. However, many countries are signatories to the European Convention on Human Rights and Biomedicine which prohibits embryo creation for research purposes. The same convention lays down criteria to protect embryos in those countries where research is allowed.

Regulation in the USA differs markedly from regulation in Europe and there are two important components. Specific laws control the funding allocated to embryonic stem cell research and others control the activities that take place. In 2001, President Bush announced a policy that limited federal funding for research on embryonic stem cell lines, restricting it to research involving only those lines that were already in existence. The corollary was that no further embryos should be developed using federal funding.

9.3.5 Procurement and control of cell lines

Whether working with bacteria, fungi, plant or animal cells, an organisation must be able to demonstrate two things to the regulators. First, the acquisition, use and disposal of the organisms must be conducted safely and ethically. Secondly, when using the cells the process must be under the operator's control. For a process to be under control the raw materials must be of predictable purity and performance – compared with more traditional raw materials, cells are of course very complex. During development the purity, asepsis and characteristics of the cells must be demonstrated. If the cells are used to express a product then, during production, these characteristics must be maintained. The product will otherwise be unpredictable and there would then not really be a process at all.

Manufacturers must be able to show that the process is being operated within a set of conditions known to keep the cells performing in the way that is intended. For all cell types the phenotype (i.e. the behavioural identity of the cell) is the result of the genotype under the influence of environmental factors. These environmental factors include all the features shown in *Figure 9.1*. These features must be controlled within acceptable ranges before there can be confidence in the quality of the product.

Figure 9.1: Environmental influences on phenotype. ▶

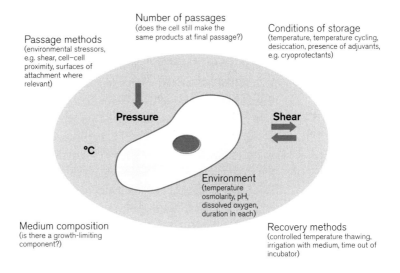

Number of passages (does the cell still make the same products at final passage?)

Passage methods (environmental stressors, e.g. shear, cell–cell proximity, surfaces of attachment where relevant)

Conditions of storage (temperature, temperature cycling, desiccation, presence of adjuvants, e.g. cryoprotectants)

Pressure

Shear

°C

Environment (temperature osmolarity, pH, dissolved oxygen, duration in each)

Medium composition (is there a growth-limiting component?)

Recovery methods (controlled temperature thawing, irrigation with medium, time out of incubator)

Early identification of those features of the process that exert most influence over cell behaviour is recommended. Once these are known the acceptable range of environmental influences can be defined. These findings can be confirmed using a systematic set of cultures with carefully controlled parameters, with the aim being to gain regulatory approval as quickly and as smoothly as possible.

Production using human cells or tissue poses its own special challenges. For the market in the USA, the regulations 21CFR1270 'Human Tissue Intended for Transplantation' and 21CFR1271 'Human Cells, Tissue and Cellular and Tissue-Based Products' will apply. Within Europe all such activity will fall under the European Directive 2004/23/EC, which gives the requirements for the sourcing, transport, holding and record-keeping of these materials. The scope of the directive covers haematopoietic cells, cord blood stem cells, bone marrow stem cells, reproductive cells, foetal cells and tissues, and embryonic stem cells. It does not, however, apply to blood and blood products (these are addressed by directives 2000/70/EC and 2002/98/EC), nor to organs for transplant.

9.3.6   Waste regulation

All manufacturing processes lead to the generation of waste. Under the European Integrated Product Policy in the UK, the environmental impact of any industrial procedure is designed carefully to include appropriate precautions that will be taken during disposal of waste products. The wastes from biological production processes may present special disposal problems because there is the opportunity for environmental damage or disease transmission if disposal is reckless. For example, human tissue must be destroyed to avoid disease transmission, and genetically modified bacteria (used for the expression of a biological product) must be deactivated before release as waste. A summary of the waste disposal regulations is available from the HSE (www.hse.gov.uk).

GMMs are closely controlled as they are one of the most frequently used tools available to biotechnology and their uncontrolled release into the wild can be very dangerous. Risk assessments are required for each stage of any industrial operation, including waste disposal, and the assessments will be scrutinised by inspectors. There must be a robust 'barrier' between the GMO and any long-term effects to the environment. If the organism is incapable of survival outside the manufacturing environment then this alone may be judged to be a sufficient barrier for low risk organisms. Any decision about disposal arrangements must be notified to the competent authority (in the UK the subject is managed between the HSE and the Environment Agency) and permission must be obtained before the decision is implemented.

In the USA waste disposal arrangements are managed under the control of the Environmental Protection Agency and operations are subject to the 'Resource Conservation and Recovery Act'. This Act requires reduction at source and safe disposal, with control of transport in between. 'Reduction at source' means that any process that will generate biological waste must be designed so that the minimum amount of hazardous waste is left at the end.

9.3.7   Health and safety regulations

Regulations also exist for the protection of the workforce. In the UK the basis for most legislation is the 'Health and Safety at Work Act 1974'. The Act applies both to employees, to visitors to the site of an organisation, and to members of the public who may be affected by activities at the organisation. In general, the basis of good safety management is to conduct a risk assessment and to decide what risks are present, what the likelihood of occurrence of any dangerous event will be, and how severe the outcome would be if it were to occur. Armed with this knowledge steps must be taken to remove, reduce or mitigate the risk. Organisations are obliged to keep all records of risk assessments and other activity relevant to health and safety management. These assessments and action plans must be used as 'working documents' and must be accessible to all. They must be kept up to date and not merely filed away and forgotten.

9.3.8 Medical products

Some of the greatest rewards in biotechnology come from its application in medicine and for this reason the regulations for medical products are considered in some detail here. Compliance with regulatory requirements (including current awareness, proper record-keeping, the acquisition of relevant authorisations, and preparation for inspections) demands a broadly similar approach across the different areas of bioscience. The regulation of medical products gives the 'flavour' of the issues involved (see *Case study 9.1*).

Case study 9.1: BoostSkin
An alternative way forward for start-up companies

Consider a hypothetical product 'BoostSkin' in which cells are used to make a skin repair product within small cartridge-type bioreactor units. The company is now about to embark on development for manufacture. What features of the regulations influence their planned activities? To minimise costs and maximise chances of success they would do well to adopt the following strategies:

- adopt and stick to well-tried technologies: it will be easier to plan and demonstrate a robust process to an inspector
- link the quality and properties of the product made at research stage to the product made at full manufacturing scale (using automation if possible)
- base the process upon a single well-characterised cell bank; this will be more attractive than to keep acquiring and using a variety of cell batches since the behaviour, characteristics and disease status of such supplies will be less predictable
- identify the critical control points in the process early on and implement the in-

process controls to govern them – this will keep product loss to a minimum
- exercise caution in deciding the product specification – identify early which features are genuinely important and avoid over-specification

Other industries (e.g. electronics) that involve a high quality assurance component have found that if the employees have a stake in the business this helps with compliance. In this quotation, ISO 9000 certification (a QMS) is being discussed by a company that had to adopt quality standards for the electronics industry: "For nearly two years before ... much effort was put in for introducing quality process management. In most organizations ... the management drives the employees hard to follow the processes. Invariably, the employees resist it due to many reasons. The [employee ownership arrangement] automatically drove the employees to be systematic in their approach and they realized that the quality processes are of paramount importance. They started following the processes without the insistence of the management"[3].

Medicines versus devices

When registering a new medical product you must decide whether the innovation qualifies as a 'medicine' ('medicinal product') or as a 'medical device'. In the EU the decision depends on the way that the material acts upon the patient. In broad terms a medicinal product is one that exerts its effects primarily through pharmacological, immunological or metabolic means. By contrast a medical device is primarily physical in its mode of action. Almost all biopharmaceuticals must, by definition, be medicines.

In the USA all medicines or devices come under Title 21 of the Code of Federal Regulations. Three categories of products are recognised: drugs, devices and biologics (e.g. vaccine or anti-toxin). The American view is that if the mode of action is 'structural' then the product is a device. If, however, the mode of action is chemical, metabolic or involves drug action then it is a medicine. Where classification of a product is uncertain then you can apply to the Office of Combination Products with a 'Request for Designation'. The regulations that apply follow from the product designation, but the key concepts of manufacturing in licensed premises, involving well-defined and controllable processes using raw materials of predictable quality, apply to all categories.

Product registration and licenses

The relevant authorisation must be obtained before a medicinal product can be placed on the market. In Europe both the product and its site of manufacture must be registered. The product is subject to a 'Marketing Authorisation' and the site must acquire a 'Manufacturing Authorisation'. In the USA there is no direct equivalent for the Manufacturing Authorisation but the authorisation to market a medicine is known as a New Drug Application. In either case the application must be made in an approved format. It is important to plan this submission extremely carefully – it is equivalent to a project in its own right.

Competent authorities and the inspectorates

In Europe the main body scrutinising medicinal products is the EMEA (this acronym is confusing as it originally stood for European Medicines Evaluation Agency, but in 2004 the organisation was renamed the European Medicines Agency – the EMEA acronym is still used, though, for continuity). Within the UK the competent authority with respect to the regulations is the Medicines and Healthcare products Regulatory Agency (MHRA). In the USA the Food and Drug Administration (FDA) regulates the markets for food, drugs, dietary supplements, veterinary products, biologics, blood products, cosmetics and medical devices. Enforcement is conducted through the Office of Regulatory Affairs.

Inspections

All manufacturers of medical goods must hold a manufacturing authorisation of some kind. Granting of the authorisation is dependent upon a satisfactory application and, usually, successful inspection of the manufacturer by the national regulatory authority who confirm that the manufacturer is complying with the regulations for good manufacturing practice (GMP). Thereafter, the regulatory authority carries out follow-up inspections at regular intervals. If unsatisfactory conditions and practices are discovered, the inspector has the power to refuse the granting of the manufacturing authorisation or to revoke an existing authorisation.

Clinical trials and small-scale manufacture

Clinical trials are used to demonstrate that a product is safe and efficacious. Following the introduction of the European Clinical Trials Directive 2001/20/EC, the preparation of medicinal products for clinical trials must now be conducted only within premises for which a 'Manufacturing Authorisation for Investigational Medicinal Products' has been issued. Once trials are complete there will be pressure to get to market so the commercial production system must be ready. The final product must bear a very close similarity to the goods tested during the R&D phase and especially to the products tested in the final stages of the clinical trials.

The obligations on the manufacturer

The manufacturer of medical products must meet certain obligations.

- To develop a product which is safe and efficacious.
- To develop a manufacturing process which will reliably produce a product which is safe and efficacious.
- To organise the human and technical resources to achieve consistency in manufacture, through systems of quality assurance (QA) incorporating GMP.
- To ensure that no product is released for sale or supply until its safety, quality and overall fitness for use has been confirmed and that the requirements of regulatory submissions have been satisfied.

Current regulatory thinking puts increased emphasis on product and process design as the key to satisfying the needs of the patient. In the USA, the FDA now uses the phrase 'Quality by Design' to emphasise this. The development group and the manufacturer must be in close contact and must actively share business objectives.

More specifically, the development process must address the key issues of product safety, ease of manufacture, identification of critical process parameters and the implementation of a comprehensive quality evaluation strategy so that regulatory approval is assured. Critical process parameters are those features of the process that exercise the most influence over the important product properties (an example of this might be oxygenation of media during a fermentation step where inadequate aeration may lead to changes in product composition).

When it comes to manufacture, the requirements of GMP must be satisfied. This means having separate and distinct production and quality units. It is a fundamental tenet of GMP that the person responsible for quality control and release decisions must be totally independent of the person responsible for production; otherwise it is conceivable that the person responsible for production might put undue pressure on the person responsible for quality control to release product purely for commercial or financial reasons. Small start-up companies sometimes struggle to achieve this separation during the early stages of business development, with a development scientist perhaps being required to wear many different hats and to set specifications, prepare material, test it and assess its suitability, without recourse to others. This may be deemed acceptable when operating under what might best be described as 'good research practice', but when material is to be prepared for use in clinical trials, the requirements of GMP come into force and the arrangement becomes unacceptable.

The quality function is normally split into two distinct groups: assurance and control.

Quality assurance. The quality assurance group has an overall managerial role, with responsibility for:

- the design, implementation and management of the documentation system, covering quality specifications, policy documents, standard operating procedures, instructions and records
- arrangements for equipment, process, system and analytical method validation (for an explanation of what this means, see below)
- staff training programmes
- vendor selection and monitoring
- ensuring stability programmes are in place, such that decisions about shelf-life and storage conditions can be justified to the regulator
- regular product quality reviews
- control of planned change
- systems for identifying, reviewing and investigating unexpected events and unplanned deviations – a typical example of this would be an unexpected trend in the microflora detected in a clean area which may indicate a breakdown in good practice
- internal audits and self-inspections against published GMP standards and company-specific policies and procedures

- a system for handling customer complaints
- arrangements for recalling product(s) in the event of patient risk

The concept of 'validation' needs further explanation. Some activities can be seen to be correct just by inspection (for example, filling a volumetric flask to a pre-calibrated mark). Others require confirmation in some way before the operator can be sure the correct result is being recorded (for example, a weighing step in which the standard practice is always to confirm balance calibration at the time using a calibration weight – this is called verification). However, some activities cannot be confirmed at the time and so are confirmed beforehand by carrying out controlled studies (the validation) to show that if they are always done in a certain way then the operator can be sure that an acceptable result will be obtained. An example of this approach is if an operator wishes to confirm that a large incubator is giving an acceptable degree of temperature control right across its height, depth and width so that any item placed within it will always be within the specified temperature of the set point. In this case a study could be conducted using the incubator and a data logger. Many thermocouples would be placed in an array within the incubator for a given time. The operator would produce a map of the temperature gradients and form a view of the match to the stated control point. These studies are often repeated at prescribed intervals in order to avoid the danger of any drift in the controls.

Quality control. The quality control unit is the department that contains the facilities for testing raw materials, intermediates and products in order to ensure conformity with specification. Some of the activity may be subcontracted. The unit is responsible for laboratory activities such as:

- agreeing sampling plans and procedures for starting materials, intermediates and products
- testing of samples and comparison of results with pre-agreed specifications
- ensuring the integrity and traceability of laboratory data
- ensuring that no raw material is released for use, nor any product released either for sale or supply for onward manufacture, until it has been tested in accordance with the registered requirements and all results have been shown to comply with registered or approved specifications

There should also be clear and written identification of those individuals and departments with responsibility for:

- engineering
- purchasing
- warehouse
- shipping and distribution

In small organisations these responsibilities may fall within the remit of the production team. To facilitate all of this the organisation must provide clear job descriptions and a chart that defines the responsibilities and reporting structures of the staff.

When working within a regulated environment, it is essential that procurement of materials and equipment are performed in a controlled manner. All chemical starting materials, including medium components, buffer components and formulation components must be purchased from designated suppliers against a pre-determined quality specification. Similarly, packaging materials should be purchased to a pre-agreed specification. On receipt, all materials must be assessed for compliance with the pre-agreed specification. In many cases, this will involve sampling from containers and testing to confirm compliance. In other cases, assessment may involve simply confirming that the appropriate grade of material has been purchased, perhaps by review of an accompanying certificate of analysis.

It is a requirement of pharmaceutical GMP that, for chemicals which comprise the final formulation of the product, every container of starting material shall be sampled and every sample analysed individually to confirm identity.

Materials initially received from vendors should be assigned 'quarantine' status and stored under secure conditions. Once the material has been confirmed as acceptable, it should be relabelled as 'passed' and only then should it be made available for use. It should continue to be stored securely under the required conditions of temperature and, where appropriate, humidity and light, and each material should be assigned a scientifically justifiable expiry date, after which it should not be used.

Equipment should be procured against an agreed 'user requirement specification' to ensure that the purchased piece of equipment is fit for purpose and meets GMP expectations. Upon receipt, equipment should be qualified and shown to be fit for use. Annex 15 of the EU GMP guide provides guidance on this issue (ec.europa.eu/enterprise/pharmaceuticals/eudralex/homev4.htm).

The primary objective of biopharmaceuticals manufacture is to produce a product that consistently meets or exceeds its quality specification, in an environment which ensures that the product is protected from contamination and is adequately contained for protection of personnel and of the environment (where necessary). The knotty problem facing the biologicals manufacturer is how to achieve all these goals, especially when some, for example product protection (which is usually achieved by passing clean air over the product to keep airborne contaminants away) and containment (which is often done by passing air from the working environment towards the goods and away from the operator e.g. in a fume hood), would appear superficially to be mutually exclusive. These apparently conflicting requirements can be reconciled and the overall quality objectives realised by careful manipulation of the following:

- air filtration
- air pressure differentials (i.e. the pressure change from one working area to another)
- physical segregation of activities
- operator training and discipline

The principle of achieving manufacturing control by manipulation of these parameters is relatively simple to understand; turning it into a practical working manufacturing environment is slightly trickier. Surprisingly, perhaps, for such an important issue in biopharmaceuticals manufacture, there is relatively little guidance on the design and operation of premises from the major regulatory bodies. For European GMP guidance, see the following Annexes on ec.europa.eu/enterprise/pharmaceuticals/eudralex/homev4.htm:

- Annex 2 – Manufacture of Biological Medicinal Products for Human Use
- Annex 5 – Manufacture of Immunological Veterinary Medicinal Products

Official US FDA requirements are largely confined to CFR 21[2].

In order to maximise product security and minimise cross-contamination risks, separate facilities should be provided for:

- cultivation and harvest (plus inactivation)
- purification
- sterile filtration of purified protein concentrate (as necessary)
- formulation and sterilisation of bulk product
- aseptic filling and finishing

It can be very confusing to follow the descriptions of the quality of the air cleanliness needed for the different manufacturing zones. In the USA, Federal Standard 209 lists the specifications as 'classes' of air quality from 1 to 100 000 in orders of magnitude. International Standard ISO 14644 also uses a 'class' designation, but the meaning is different and the numbers run as integers from 1 to 9. The EU GMP guide used four 'grades' A to D. In each case the lower the figure or letter, the cleaner the environment – a useful comparison of these specifications is available (www.s2c2.co.uk/docs/ClassificationOfCleanrooms2005.pdf). Aseptic filling must be performed in an environment which minimises the potential for particulate and microbiological contamination and so is carried out in a Grade A (Class 100) environment, often surrounded by Grade B (Class 10 000). Additionally, controlled environments are required for the washing, wrapping and ultimate sterilisation of the equipment and components to be used in manufacture and for the sterile filtration of the bulk formulated solution, where appropriate.

The constraints, controls and equipment required for such operations are very expensive to put in place and maintain. It is common for start-up biopharmaceutical companies to concentrate on what they do best – the molecular genetics, cell cultivation and biochemical separations – and contract out 'pharmaceutical finishing' to an established sterile products manufacturer.

It should now be apparent that all phases of biopharmaceuticals manufacture are performed in controlled environments and that these environments must be subject to regular and repeated monitoring in order to demonstrate maintenance of acceptable clean conditions for manufacture. This ongoing monitoring, allied to periodic re-validation of controlled environments, creates a strong demand for an on-site microbiological quality control function to sample environments, incubate samples, enumerate and where necessary identify isolates, and provide data to demonstrate control. As a log of results is built up the values are also examined for trends so as to alert the management in advance of any remedial action that may be needed.

Similarly, it is essential that all equipment associated with the achievement of controlled environments be subject to initial qualification (to ensure that it is fit for purpose) and ongoing maintenance and calibration (to ensure that it continues to be fit for purpose). This essential GMP requirement creates a strong demand for an in-house, professional engineering resource.

Process validation

It is absolutely essential that processes can be relied upon to consistently yield the expected outcome within acceptable limits. Demonstrating this reliability is known as validation. It is an essential GMP requirement that all critical steps in processing, and all changes to processes, be validated. Similarly, analytical methods used in quality control must be shown to be reliable, accurate and precise – this is known as analytical method validation. Thus, resources and responsibilities for validation activities must be clearly identified. The time, resource and cost requirements for validation must be planned carefully.

Comprehensive, accurate records must be made at the time of the activity and must confirm, via a signature, who carried out the activity. Records must provide a full, traceable history of any activity and, very importantly, must identify any planned or unplanned departure from the prescribed method. Identification and recording of so-called 'deviations' is a critical activity in quality management and there must be defined procedures to describe how such events will be recorded, investigated and their impact upon the suitability for use of the related product(s) decided upon.

All documents must be formally approved for use and must be subject to 'change control'. This means that when a draft document is officially approved for use the issue (electronic or paper) is undertaken in a controlled manner so that any copies in use in the workplace are the version that is intended for use. It is important to withdraw all previous versions so as to leave no ambiguity about which procedure is to be followed at any one time. In this way out of date instructions and documents are superseded by up-to-date replacements.

Batch records (records of manufacture and control, including deviation reports, environmental monitoring excursions, sterilisation records and other relevant data) must be thoroughly reviewed by the quality department prior to release of material. The records must be stored and retained for a defined period to permit investigation of potential problems following, for example, adverse reactions or customer complaints for that or related batches. The minimum acceptable retention times for records are defined in the GMP guides.

No batch of starting material should be released for use in production, no key intermediate or bulk drug substance should be released for further processing, and no finished product should be released for sale or supply (for example, for use in a clinical trial), until it has been shown to comply with its approved specification and to be fit for use. The quality control unit generally has the responsibility to release materials but, in the case of finished products, the European Union requires that release be preceded by formal 'certification' of the batch by a 'qualified person'. The qualified person is named on the Manufacturing Authorisation of the organisation and has the legal duty to ensure, for each batch of finished product, that:

- the material has been manufactured in accordance with national law and the provisions of the relevant authorisations
- all the necessary tests have been performed to show compliance with the marketing authorisation (or product specification file in the case of investigational medicinal products)
- the material has been manufactured in accordance with the principles of GMP as defined in the European GMP directive (2003/94/EC)
- material imported from a third country (outside the EU or European Economic Area) has undergone the necessary testing to confirm compliance with the marketing authorisation
- certification is made by the qualified person, in a register or equivalent, that all the above requirements have been met

Then, and only then, can the product be released for sale or supply.

European law lays down strict eligibility rules for people wishing to act as a qualified person. The educational and experience requirements are stringent and, consequently, only a limited number of people are eligible to act in this capacity. Nonetheless, every organisation holding a Manufacturing Authorisation must have at their disposal the services of one, or more, qualified person(s). Recruiting and retaining a qualified person will represent a significant challenge to a small, start-up biopharmaceutical company, as they tend to command high salaries and few of them have a thorough understanding of the specific quality and patient safety issues associated with the very specialised activity of biopharmaceuticals manufacture. To resolve this problem a common approach is to employ an experienced quality assurance manager, to train the staff carefully in the regulatory requirements, and to engage the services of a contract qualified person, named on the Manufacturing Authorisation, for a certain number of days per year.

9.4 – The impact of regulation on business

9.4.1 Pilot facilities

Costs for creating and accrediting a facility to GMP are considerable. A practical alternative is to avoid the capital cost and to seek a contractor who already has the capital equipment installed and for whom the validation of the process will be the main task rather than design-and-build. Contract costs to provide a basic aseptic product to a validated protocol are usually in the range £1–10 millions of pounds (depending on process complexity). However, this may still be the preferred approach if the new company is inexperienced in process development.

9.4.2 Where the money goes

A manufacturer involved in the preparation of, for example, a biopharmaceutical for early-stage clinical studies will require small scale manufacturing facilities, will operate a relatively simple QMS and use simple GMP systems to support infrequent, small scale manufacture. They may retain the services of a qualified person on a contract basis, or utilise the services of the qualified person of a contract manufacturer or packer. However, once the manufacturer proceeds to large, multi-site clinical trials, the full panoply of GMP expectations arise. Thus, in addition to the cost of supporting a multi-site, perhaps multi-national, clinical trial, the organisation must invest in large scale manufacturing facilities operating to full GMP. The facilities must be supported by a sophisticated QMS and will be subject to frequent, expensive regulatory inspections.

Such costs would be prohibitive to many small start-up companies and it is no surprise to find that very few such companies have made the transition to 'big pharma' status although Amgen, Genentech and Biogen Idec are three success stories that spring to mind. It is more common for small start-up companies to have a strategy of developing products up to 'proof of principle' (Phase II Clinical Trials) and then either joint venturing with 'big pharma' to offset the costs of further development, or simply selling or licensing the rights to the product to a large pharmaceutical company (see *Chapter 6* for

Case study 9.2: Synergen
Insufficient cash to support product development

In the 1990s, a US start-up company called Synergen developed a product for the treatment of inflammatory disease. The product was a recombinant analogue of the human receptor protein for interleukin 1 (IL-1) and was produced in *Escherichia coli*. Synergen built a very large and impressive manufacturing facility in Boulder, Colorado, to produce the protein for use in clinical trials. Early results were encouraging but, unfortunately, results from Phase II 'proof of principle' studies were not as good as had been expected. The Synergen investors got cold feet and the company collapsed.

However, some people saw the promise in the product and persuaded Amgen to buy what was left of Synergen and the rights to the product. Amgen were prepared to support the product through further clinical trials and now, under the name Kineret®, this product is a successful treatment for inflammatory diseases such as rheumatoid arthritis.

Gail Naughton
Pioneer in Tissue Engineering

The career of Dr Naughton illustrates the importance of regulatory compliance. As a co-founder of the pioneering company Advanced Tissue Sciences, Inc. (ATS) her experience underlines the importance of gaining fast, smooth regulatory approval in securing the future of a business. In 1987 she co-founded ATS, an organisation which built on some of her own patents that described growing human tissue on synthetic scaffolds for repair of the body. Over the next few years regulatory affairs were to have a major influence over business viability. In the early stages there was a lot of negative feedback about the product. Burns surgeons did not believe that the first product, TransCyte™ (a non-living skin replacement for use in burns treatment), would receive FDA approval as the whole concept was so radically new. This did not deter her and, with persistence, approval was eventually granted. When Dermagraft™, the second product (a cryopreserved living dermal tissue), was developed it encountered regulatory delays. Dermagraft™ was intended for use in acceleration of healing of chronic ulcers of the foot. The FDA requested additional clinical trials which resulted in a 3-year delay in the launch of the product. Confidence in the company's future fell and the stock price fell with it.

The company's response was to refine the processes using robotics, to apply more robust process controls to reduce the incidence of any batches that fell outside specification and to introduce efficiency measures to reduce manufacturing costs. The slimmed-down company was eventually purchased by the Smith & Nephew group at which time Dr Naughton remained vice-chairman while taking up a post as Dean of the College of Business Administration at the University of California, San Diego. Dr Naughton thus became the first female biotech entrepreneur to lead a major business programme at a US university. More recently the company has again changed hands when it was sold to Advanced BioHealing Inc. of New York. Sales of Dermagraft™ resumed in February 2007.

Dr Naughton's philosophy is one of the importance of self belief "you can accomplish the impossible, but sometimes it takes a little longer than you thought".

further consideration of the alternative strategies that may be adopted by the start-up company at this stage). The larger company can then allocate the massive resources required to carry out pivotal clinical studies and will be able to withstand the financial consequences if the product fails at Phase III.

Start-up biopharmaceutical companies are often described as having a high 'cash burn' (high running costs associated with premises, R&D, etc.), but these costs are small compared with what is required to support a multi-national clinical trial. With no revenue coming into the company it is often a race to get to 'proof of principle' and attract the interest of 'big pharma' before the cash invested runs out. From time to time good companies, with potentially successful products, go to the wall simply because their financial backers decide that enough is enough (see *Case study 9.2*).

9.5 – References

1. Guidelines for Research Involving Recombinant DNA Molecules. (1980) *Federal Register*, 45: 6724–6749.
2. Physical establishment, equipment, animals, and care. (2006) *Code of Federal Regulations*, Title 21, Volume 7, Section 600.11 (available from the FDA website at www.fda.gov).
3. **Prasad K** (2003) Running a virtual enterprise. *Ubiquity,* **4** (8).

9.6 – Additional resources

A Step by Step guide to COSHH assessment. (2004) HMSO, London.

Health and Safety Executive (1992) *A Guide to the Genetically Modified Organisms (Contained use) Regulations.* HMSO, London.

Rules and Guidance for Pharmaceutical Manufacturers and Distributors. (2007) Pharmaceutical Press, London.

www.hse.gov.uk/pubns/hsc13.pdf

Health and safety regulation ... a short guide, Health and Safety Executive,

Learning outcomes

Key learning points from this chapter are:

- that the regulations that surround bioinnovation are there to protect customers, employees and the pubic and to promote best practice

- that if the regulations are embraced actively, they can become a tool in the search for competitive excellence

- it is of vital importance to think creatively about the ways that the regulations may be satisfied and to stay ahead of regulatory developments through personal reading and professional updates

- in innovation-based industries the biggest share is gained by the company that is first to market; any

organisation that thinks ahead, conducts the research and development with careful consideration of robust manufacture, and links the methods used in its early findings with the methods used to make the final goods, is likely to receive approval much more quickly that one in which work must be repeated

- it is much easier to structure a company in such a way that the procedures match the regulatory standard than to build a company on other principles and then try to meet the regulatory requirements afterwards

Chapter 10
Ethical Issues

Rob Lawlor

10.1 – Introduction

Section 10.2 considers the ethical issues raised by each of the previous chapters. In some cases, these discussions will be rather superficial, in others, the ethical issues involved will be identified but not discussed in detail. Nevertheless, this chapter will be beneficial for two reasons:

- For those who feel uncertain about what is meant by the term 'ethics', this short introduction will provide a better understanding of what ethics is and what ethics involves than any definition could
- the discussion of these various issues will highlight the breadth of ethics, demonstrating that ethical issues arise (almost) everywhere

Of course, some ethical issues are more important than others, and some are more complex than others. If you are playing Monopoly with a friend, and he leaves the room to get a drink, you know that that you should not cheat by taking a few thousand pounds and a couple of hotels while he isn't looking. This is not a complex case and neither is it an important one. Nevertheless, asking whether or not one should cheat *is* an ethical question, albeit a fairly unimportant one. The ethical issues addressed in this chapter, however, are somewhat more important and more complicated than the one stated here.

The next section of the chapter considers the conflict between people's needs and the protection of IP, and this is followed by an exploration of the ethical issues that arise when benefits are weighed against risks. The final section looks at the ethics of individual choices, concentrating primarily on a line of argument that is frequently used as an attempt to justify what seems to be an ethically dubious decision: 'if I don't do it, someone else will'.

10.2 – Summary of ethical issues

Knowledge and technology transfer

In *Chapter 1*, risk was discussed. In relation to risk, it is important to distinguish between economic risk and risks to health and safety, and also the need to distinguish between taking risks (oneself) and imposing risk on others. Some would argue that entrepreneurship simply involves taking economic risks, and that this isn't an ethical issue. Risk, they might argue, only becomes an ethical issue when risk is imposed on others, for example, by putting their lives at risk. While this is certainly an important ethical issue, it is not true that there are no ethical issues relating to the taking of economic risks. Whenever your actions will affect others, for better or for worse, there is an ethical dimension. Unless your business is a one-man-band your decisions *will* affect others, as shown by the authors of *Chapter 1*: "being a risky and often unstable commercial sector, the commercialisation of technology presents issues regarding the well-being and financial security of employees. Failure of spinout companies obviously affects people who work for them – and there is a clear need for managerial responsibility in considering this in initiating and managing projects." The issues raised by imposing risks on others will be discussed later in the chapter.

Generation of ideas

In *Chapter 2*, the authors discussed the generation of ideas. Many readers will doubt that there are ethical issues associated with the mere generation of ideas. But, of course, this is not about the generation of ideas just for the sake of it; rather, this is about generating ideas that will be acted upon and put into practice. Ideas generated specifically at the very beginning of an enterprise (ideas about the enterprise itself) offer a major opportunity for a business to be a pioneer in ethics as well as in enterprise (see *Section 10.5* for a brief discussion on Fairtrade and Ecover). Ultimately, ethics is not just a matter of following rules and abiding by regulations; beyond this, ethics can require imagination and the generation of ideas. A number of examples of pioneering ethical thought in enterprise are discussed in Schaper[1].

Protecting ideas

Protection of intellectual property was covered in *Chapter 3* and most readers will recognise that this is clearly an ethical issue. This is considered in greater detail in *Section 10.3*. For a detailed discussion of issues such as the patenting of genes, please see *Section 10.7* – Additional resources.

Researching ideas

Chapter 4 covered identifying the competition and analysing the strengths and weaknesses of competitors. On the face of it, the mere collection of information may not seem to be an ethical issue, but it is. First, there are methods of collecting information that are clearly unethical (breaking and entering, tapping phones, etc.) and others that are ethically questionable (searching through rubbish or 'infiltrating' other companies). Secondly, this information is not gathered for the sake of it. Rather, it is accumulated to inform decisions and to develop strategies. Clearly there are ethical issues associated with this. As an example, consider predatory pricing. The management of a company may believe that it has an advantage in relation to its competitors in that it can afford to run at a loss for a period of time, while its competitors cannot. Thus, prices are dropped so low that the company runs at a loss knowing that its competitors must also drop their prices if they are to remain competitive. The competitors may then go out of business allowing the company to raise its prices, making a bigger profit in the longer term, with less competition. It is not at all obvious that larger companies should be permitted to act in this way.

Communication and funding of ideas

These inter-related topics can lead to the question: 'where do we draw the line between positive spin and outright deception?'. For example, imagine that you have a meeting with a potential backer, but you suspect that the backer would not fund your project if he knew a particular piece of information. You decide that, if he asks, you will be honest, but you tell yourself 'if he doesn't bring it up, why should I?'. Is this acceptable? You might tell yourself that you haven't told a lie, so you haven't done anything wrong. But can you say this sincerely, or is this just a bit of sophistry? Would your judgement be the same if you were on the receiving end of this 'deception'? Of course, whether or not we consider this to be deception will often depend on the details, but situations do arise where omissions of this sort are clearly misleading. Crane and Matten[2], for example, state that in the 1980s and 1990s some companies claimed their products were biodegradable but failed to mention that "biodegradation was only actually possible under highly unlikely conditions".

Setting up a company

Setting up a company involves a number of ethical issues, ranging from the nature of the business itself (for example, is it ethical to start a business producing and exporting weapons?) to the everyday running of the business, including dealing with employer–employee relations (for example, what is a fair wage for your employees?). The ethical issues associated with the nature of the business are discussed in *Section 10.5*, but the issues related to the everyday running of the business are beyond the scope of this chapter (*Section 10.7* – Additional resources, includes details of where to look for further information).

The business plan

The role of the business plan was covered in *Chapter 7*. The range of ethical issues associated with the drawing-up of a business plan is diverse and includes many of the topics discussed above.

Regulations

It is important to recognise the crucial role of regulations in the biosciences, but one should also be aware of their limitations. Ethics isn't *just* about following regulations and adhering to professional codes of conduct as, hopefully, this chapter will demonstrate. However, following regulations is a *part* of acting ethically. Regulations can be seen as minimal requirements, but this isn't the only way in which regulations can be important (see *Case study 10.1*).

Case study 10.1: The tragedy of the commons

Hardin[3] describes a situation in which a number of herdsmen graze their cows on a common pasture. Individually, it is in the interests of each herdsman to graze an extra cow on the common. So, if a herdsman puts an extra cow on the pasture (but no one else does), he will be better off than he was before. The more cows the herdsman can graze on the common and sell later for a profit, the greater his profit will be. Adding another cow will obviously contribute to overgrazing of the common. This negative result must be considered too, but on a common pasture, all the herdsmen share this cost, while the profit is all his:

"... *the rational herdsman concludes that the only sensible course for him to pursue is to add another animal to the herd. And another... But*

this is the conclusion reached by every rational herdsman sharing a commons. Therein is the tragedy ... Freedom in a commons brings ruin to all."

The only solution, Hardin argues, is "mutual coercion, mutually agreed upon by the majority of the people affected". Appealing to people's conscience or their sense of responsibility would only penalise the conscientious while doing little to solve the problem. As an individual herdsman you may refrain from putting another animal on the commons but, if others don't refrain, the commons will be ruined anyway, and your restraint will have been futile. Leaving the fate of the commons to individual choice is likely to be both unfair and ineffective. The only solution that will be both fair and effective will involve the use of coercion.

Of course, this analysis isn't limited to the grazing of animals. A similar analysis can also be applied to fishing, pollution and, most significantly, global warming. Suppose, for example, that you run a business and want to cut the emission of pollutants, but you have concluded that it would be very difficult to remain competitive if you did. In these circumstances, it is not necessarily hypocritical to decide against cutting emissions while also campaigning for stricter laws that would *force* you to cut your emissions. Crucially though, the regulations would also force your competitors to cut *their* emissions too. As such, regulations needn't force people to do what they don't want to do. Regulations can also be used to ensure that we all do what we would want to do anyway – if only we could trust others to act in the same way.

10.3 – Intellectual property

Chapter 3 argues that without the protection afforded by IPR, pharmaceutical companies would be unlikely to develop new drugs as they need the monopoly provided by IPR to recoup the money spent developing each drug. This section discusses this and other arguments for and against IPR in detail.

Consider *Case study 10.2* which demonstrates why many people are keen to argue that IPR cannot be justified, but also reiterates the point made above.

Case study 10.2: HIV treatment

HIV is currently expected to devastate large swathes of the black population, not only in South Africa, but also in most parts of the continent. In 2001, 70 per cent of the world's 40 million people infected with AIDS lived in Africa, as did 2.3 million out of the 3 million who had died from the disease. In South Africa, where every ninth person is HIV positive, life expectancy is now [2004] a mere 47 years – without AIDS it would be 66. Even worse is the situation in other Southern African countries such as Mozambique or Swaziland, where the average life expectancy is well below 40. If things go on as they do, 7–10 million South Africans will have died of HIV/AIDS by 2010.

With recent progresses in pharmaceutical research, however, there is the theoretical chance of treating HIV-infected people. The HAART (highly active anti-retroviral therapy) medication cocktails calm down the symptoms of the disease and lower infection rates, especially for mothers giving birth to children. The only problem is that such therapy would cost between €10000 and €15000 per person per year – a sum completely out of the reach of governments that can hardly spend €10 on the health of their citizens each year.

When the issues entered the political agenda in the mid-1990s, the pharmaceutical companies were reiterating the fact that they had to cover their enormous costs for developing the medications. With development costs of €800m for each new drug released on to the market, and an average 12 years' research time necessary to bring the drug from inception, the drug companies might be said to have quite a strong case for wanting to reap the benefits of their investment. Conversely, who else but these companies is in a position to provide help for these appalling conditions in the other half of the globe?

Taken from Crane and Matten[2], *Business Ethics*, pp. 107–108.

Broadly speaking, there are two common approaches that people adopt in order to defend IPR. The first is to appeal to the consequences of different policies and to argue that the consequences of not protecting IP would be even worse, and that it is this fact that justifies IPR. The second approach, on the other hand, doesn't appeal to the consequences of different policies, but appeals directly to rights – the rights we are trying to justify are *legal* rights and the rights we are appealing to, in order to justify these legal rights, are *moral* rights.

10.3.1 The consequences of alternative policies

Pharmaceutical companies use this approach to justify IPR and the prices of their products. Essentially, the idea is that without these rights the pharmaceutical companies would not be able to make a profit. This would mean that there would be no incentive to produce the drugs in the first place, in which case *no one* would have access to treatment. The situation would therefore be significantly *worse* overall.

As suggested in *Case study 10.2* there is clearly some truth in this but the argument has one significant flaw: it considers only two options, either the status quo or a complete absence of IP protection. If these were the only two options available, then the argument presented might be a strong one. But, of course, there is no reason to think that these are the only options. One alternative would be to provide more limited protection. For example, the IPR could be used to protect companies in certain 'rich' markets, but not to prevent other companies from selling identical drugs in poorer countries and at significantly lower prices. Alternatively, the protection conferred by IPR could be combined with a requirement that the company holding IPR should provide the drugs in poorer countries at a low price that simply covers, for example, the costs of production and distribution.

An appeal to the consequences of alternative policies may therefore justify IPR in general, but may not justify the particular IP laws that we have at the moment, and may not justify the sort of IPR that large pharmaceutical companies are keen to defend. However, there are reasons for thinking that a move towards IPR with particular requirements, such as differential pricing to allow provision for the most needy, may not be straightforward. In the mid-1990s, GlaxoWellcome developed a drug against malaria, called Malarone. GlaxoWellcome recognised that they were not likely to make much money selling the product in Africa and decided to make the drug freely available in African countries[2]. Unfortunately, however, there were problems with distribution of the product. GlaxoWellcome had planned to distribute 1 million doses within 3 years, but only managed to give away about 100 doses. Crane and Matten[2] suggest that there were a number of reasons for the failure of this initiative:

- the health authorities in African countries were suspicious, suspecting, for example, that the donation programme might be an experiment
- there was the problem of deciding a just and efficient way of distributing the drugs because, even if all of the doses could be distributed, there were not enough for everyone who needed the drug
- there was concern about the possibility that supplies of the drug might be stolen and sold on the black market

These were problems the company faced when trying to give the drug away, voluntarily. However, it is reasonable to suggest that the situation would be different if something like this was required by law. Certainly, with regard to the issue of African health authorities' being suspicious of the company's motives, it is unlikely that this would be such a significant problem if the African countries recognised that the company was required by law to provide the drugs 'at cost' to poorer nations. Further, presumably the problem of deciding how to distribute the drug would not normally be a significant impediment as even an imperfect distribution system would be better than none.

However, the final concern, over the fear of a black market, would be the most likely stumbling block. This is particularly the case if differential pricing is being considered, so that the same drug is available in one country at one price but available in other countries at *much* higher prices. Under these circumstances, the pharmaceutical companies would clearly be concerned that the cheaper drugs would find their way on to the black market in the wealthier countries. Nevertheless, it would seem unduly pessimistic to think that nothing can be done and that we should simply accept the status quo. It is important to emphasize that we are attempting to defend IPR here solely by appeal to the consequences of alternative policies. It seems unlikely that the best consequences for *all* concerned will be brought about within the framework of current laws, as opposed to an alternative with some requirements for the inclusion of differential pricing. A law requiring differential pricing in order to help provide drugs to the poor may result in some black market activity and this may mean that some pharmaceutical companies don't make as much profit as they would otherwise. But if we are considering the wider consequences, including the benefits to the poor – and if we are appealing to these consequences to justify IPR, without any appeal to authorship rights – it seems plausible to think that the consequences of implementing these new laws would be better than the consequences of the existing laws.

Even if it is not required by law, individual entrepreneurs can of course *choose* to implement differential pricing, and to work to help the poor while also making a considerable profit.

10.3.2   The rights-based approach

As an alternative to the approach considered above, we can try to justify IPR by appealing to rights directly. According to this view, the 'author' of an idea has a right to the idea, and to any profits made from the idea, simply by virtue of the fact that he is the author: it is *his* idea and his rights of ownership ought to be protected. Some people seem to be suspicious of this approach, claiming that while it is obviously true that one can own physical objects, one cannot own ideas. Similarly, some also claim that the idea of IP is a new concept that has been introduced by people like software developers or pharmaceutical companies who just want to protect their profits.

On the contrary, the claim that people have a right of some sort to their ideas, and that this right should be protected, is actually fairly intuitive. Suppose Joe and Jimmy work for a marketing company and they have both been given the same assignment: to develop ideas for the marketing of the company's next big product. Just before a meeting with senior management, Joe asks Jimmy, "any good ideas?". Joe and Jimmy are friends and so Jimmy trusts Joe and happily explains his idea. Joe realises that Jimmy's

George Koukis

George Koukis was born to illiterate parents in a small town in Greece in 1946, one of four children. George states, "We knew we had no money so we learnt not to ask for anything". However, he did receive daily encouragement and was taught about "integrity, doing the right thing, caring about people and the community". George's father wanted him to go out to work but his mother wanted him to study in order to have a better life and, as George says, "mother won".

After military service George worked as an accountant but, believing that Greece didn't offer great opportunities, George left Greece, moving to Belgium and then to London. Unable to get a work permit, he cleaned toilets, washed dishes and worked in pubs until his application to immigrate into Australia was accepted. While in Sydney, in the 1970s, he worked from 9am to 5pm as a clerk, attended university from 6pm to 9pm, and then worked as a waiter at night. Studying was squeezed in late at night or at weekends. This continued for seven and a half years.

George worked for Qantas and then, in the eighties, for the Australian arm of Management Science America. In 1987, while still in Sydney, George started a venture in share trading and within 3 months was making "obscene money". However, in October 1987 the stock market crashed and he lost everything in one evening. George recalls that, while others might have committed suicide, he reminded himself of his grandmother's adage, "Every obstacle is for your own good", and he worked hard to regain the things he had lost.

In 1993, George bought COS Software Engineering, a company based in Switzerland, for $948000, using venture capital he obtained in Hong Kong; he gave the venture capitalist a 50% share in the company. When he bought it, the company had two offices, one in London and one in Geneva, 19 small clients, small revenues and was losing millions of pounds each year. George made Geneva the headquarters of the company and renamed it TEMENOS. To the dismay of his venture capitalist, George didn't give his new team any financial objectives. He gave qualitative objectives only.

In 2000, George repaid the original loan and bought out the venture capitalist for $95 million. On 26 June 2001, the company was listed on the Zurich main Board with a value of about $800 million – today the market value of the company is $1.2 billion. In June 2006, an independent US research company listed TEMENOS as the number one company in the world amongst those companies providing computer systems for banks.

The TEMENOS systems cost banks anything between £1 million and £30 million. It was obvious to George that banks in developing or poorer countries could not afford these prices. TEMENOS therefore took a version of their top-selling system, packaged it differently, made it very easy to install and made it available to non-governmental organizations (NGOs) at a cost of a few thousand pounds. These NGOs then provide people in developing countries with personal loans so that, for example, they can set up a food stall or a small farm. TEMENOS also donate their software, free of charge, to universities around the world so that students can have hands-on experience of learning how to use new technology and modern designs.

George offers the following advice:

- Do not feel jealous about what other people have or their achievements. If it is positive then emulate it, go after it yourself.
- Do not harm anyone in any way, be it stealing, destroying or abusing. Be positive.
- Do not be afraid of failure. If you fall, dust yourself off and keep going.
- Profitability is the result of having happy and satisfied clients – not the difference between revenues and expenses! Have a longer horizon in mind when you build a business. Do not be tempted with becoming rich quickly.
- A happy employee rarely leaves the company. Take care of the most precious asset you have.
- Once I was told that I can either build an ethical company or a successful company, so I proved that I can build an ethical and successful company.
- "The fish rots from the head": if the head goes you cannot eat the body; metaphorically speaking, if the head of any unit is not ethical then the whole structure will fall.

idea is much better than his own. He decides that he will make sure he gives his presentation first and, instead of presenting his own weak idea, he presents Jimmy's instead. In this case, it seems clear that Joe has acted unethically. Not only has he stolen Jimmy's idea, but as a result of this he also receives the esteem, the credit and, let's say, the promotion that should rightfully be Jimmy's. The important point to note is that these conclusions do not depend on considerations of consequences. Joe's actions are wrong simply because he has taken the credit for something that is rightfully Jimmy's.

Similarly, consider the following example, which also illustrates that IPR is not a new concept. The novelist Helen DeWitt writes and publishes a new novel using her own money. The book becomes a bestseller and makes a significant profit. Who should get the proceeds? The obvious answer is Helen DeWitt. Who else would have a claim to any part of the profit? On the face of it, at least, it seems that Helen DeWitt has an entitlement to the proceeds made from selling her novel simply in virtue of the fact that it is hers – she wrote it. Again, the important point is that this example does not depend on the consequences. Helen DeWitt gets to keep the profits not because this somehow maximises good consequences, but simply because she is the author. Furthermore, she can do what she likes with the book. If she wanted, she could keep the manuscript locked in a drawer, depriving the world of a great novel.

Why is this example illuminating? First, because it highlights again the fact that we intuitively think that authors should receive credit and payment for their work (in contrast to the views of those who are suspicious of the idea of IP, and think you cannot own ideas). Secondly, it highlights the point that this conclusion does not depend on the consequences of different policies. Thirdly, the rights that authors have are not limited to rights to profits. Authors, for example, also have a right to keep their work to themselves. Finally, this example is illuminating because Helen DeWitt doesn't merely own the books she produced. If you were to copy out the text of her novel and then produce your own copies of the book, under your own name, this would actually be a more serious crime than if you merely stole a physical copy of one of her books. Furthermore, she doesn't merely own the sequence of words. She owns something more abstract than that. If, for example, you were to translate the novel into Spanish and then publish it under your own name, this would be just as great a violation of Helen DeWitt's rights.

If we now consider IPR in relation to the marketing of drugs rather than literature, we may be less confident in the conclusion that IPR should give people the right to do whatever they want with their product. In contrast to the novelist who decides that no one should have access to her novel, consider a pharmaceuticals company that decides not to make a new wonder drug available to the public. Suppose, for example, that the company has another drug that can be sold instead at a much higher price. For an alleged example, see Boseley[4]. While we might accept that a novelist has the right to keep the fruits of her labour locked away, we are unlikely to condone similar behaviour on the part of the pharmaceuticals company.

10.3.3   Moral pluralism: appealing to rights and consequences

On the face of it, the above example from 'Big Pharma' might seem to offer a strong objection to the rights-based approach. However, more can be said in defence of rights. First, although we should not dismiss our concerns regarding the behaviour of a pharmaceutical company inclined to withhold a wonder drug for its own selfish reasons, neither should we dismiss our earlier thought that there is something right about the idea that the author/inventor has a claim that others do not. Secondly, rights need not be seen as absolute. The fact that someone has a particular right need not be a conclusive consideration. Rather, it may be a *pro tanto* consideration: a consideration that has some weight but may not be conclusive. According to this pluralist approach, rights matter: not instead of consequences, but as well as consequences. Thus, on this account, we can grant that pharmaceutical companies have *moral* rights to their IP but go on to conclude that, when defining legal rights, we should consider the consequences of different policies. As well as protecting the rights of ownership, IPR laws should also provide some form of provision for the poor and needy.

10.3.4   Final note: protecting the weak

Frequently, when discussing the ethics of IP, people tend to present it as a conflict between the interests of big business and the needs of the poor. Until now the arguments put forward in this chapter have been presented in a similar vein. However, it would be wrong to suggest that IPR invariably favour the rich over the poor. IPR are largely blind to wealth and poverty in that they simply recognise authorship. If you translate Helen DeWitt's novel into Spanish, and make a fortune marketing it as your own work, you violate her rights whether she is rich or poor. As such, IPR can also protect the poor, and can protect the independent entrepreneur.

In 1993, Chiao[5] discussed the problem of IP laws not being consistently or rigorously enforced in China and gave the following example:

"In 1985 [Jin] discovered that a colleague had sold a factory in southern China his patented idea for making use of atactic polypropylene, a chemical by-product. But his attempt to seek retribution was foiled by another common problem – a reluctance to sue. University officials told him only

that the action should be 'criticised'. 'They even implied that I should be flattered by the rip-off', Jin said."

Clearly, IPR is not always about protecting the profits of large companies. In this instance, if IP laws were enforced rigorously, they would be used to protect the interests of an individual researcher. In contrast, the example above illustrates what can happen if IPR laws are not enforced and rights are not protected.

Isaak[6] described *The Honey Bee Network*, a newsletter published in six languages, which aims to protect the IPR of inventors in developing countries, as follows:

"The Honey Bee philosophy is aimed at remedying... the temptation of large corporations from developed economies to take or buy out the innovative technologies of indigenous people in developing countries, without giving them sufficient credit for their intellectual discoveries or payment for their ideas and inventions."

Again, this is not an example of IPR being used to protect the profits of large companies. On the contrary, this is an example of IPR being used to protect the interests of inventors in developing countries, and in particular to defend them *against* large companies, protecting them from exploitation.

As described earlier, IPR are largely blind to wealth and poverty. The point is not to help the rich or the poor, but rather to recognise *authorship*.

10.4 – Risks and benefits

Frequently, new technologies promise to revolutionise our lives, providing all kinds of benefits. Karl Popper[7], for example, talked about major changes in people's lives following the introduction of the steam engine and then by the "Henry Ford revolution, which made motor cars available not only to millionaires but to workers". Popper claimed, "These are the great revolutions, and they cannot be foreseen by anybody... And nobody can predict what will be the next great invention in terms of personal services." Frequently, however, new technologies bring with them potential problems. The motor-car, for example, has made a major contribution to pollution and global warming and has been responsible for a great number of deaths through accidents.

In short, new technologies can provide benefits but these are frequently accompanied by risks. Nanotechnology is a notable and current example of this. Nanotechnology promises to revolutionise our lives in innumerable ways but, at the same time, the public has concerns about the safety of nanotechnology.

This raises two questions. First, how should we deal with the risks associated with new (and existing) technologies? Secondly, how should we respond to the public's concerns? A common response to the second question is simply to dismiss public concerns. Most members of the public don't understand the science associated with sophisticated technologies and, consequently, it is tempting to think that we should therefore dismiss public opinion as uninformed and invalid. Having dismissed these 'unfounded concerns', a common response to the potential risks involved is to carry out a risk–benefit analysis. That is, the risks are calculated and weighed against potential benefits. The remainder of this section considers some of the problems and ethical issues associated with these responses.

10.4.1 Why we shouldn't dismiss public concerns

This argument is probably best expressed in the form of an analogy. Consider the doctor/patient relationship. In medical ethics the 'doctor knows best' model is often contrasted with the 'informed consent' model.

Risk–benefit analysis – the 'doctor knows best' model

According to the 'doctor knows best' model, the patient is simply not qualified to make a decision about treatment – they do not have a sufficient understanding of medicine, or of their illness. The doctor is therefore justified in prescribing the appropriate treatment without any real dialogue with the patient.

However, the complaint is that, with this approach, the doctor doesn't give enough weight to patient autonomy and doesn't treat the patient with sufficient respect. In contrast to the 'doctor knows best' model, the law requires a doctor to obtain consent from the patient before embarking on a particular treatment regime (except in unusual cases). It is the doctor's duty to explain, in terms the patient can understand, the potential benefits of drugs or other treatments available, and the possible risks associated with these. The patient can then decide whether or not to consent to a particular form of treatment. It is the patient's right to reject his doctor's advice. The doctor may disagree with the patient's choice but, if the patient is a competent adult, the doctor must respect his or her wishes. A doctor who treated a patient without consent could be charged with assault – even if claiming to have acted in the patient's best interests.

We can now draw an analogy between the behaviour of doctors and the behaviour of scientists tasked with the introduction of a new technology. Of course, the analogy is not perfect, but the view that scientists (or policy makers) should make decisions on the basis of risk–benefit analysis, ignoring public concerns, is clearly closer to the 'doctor knows best' model than to the 'informed consent' model.

Imagine, for example, that you have invented a groundbreaking new sun cream that only needs to be applied once and is guaranteed to protect you from skin cancer. There is a possibility that some people will react badly to the preparation and that this will result in permanent scarring. However, this sort of reaction is likely to occur only *very* rarely. You may decide that the benefits outweigh the risks and put the product on the market without making the risks known. Clearly, this is unacceptable. Such a decision is not yours to make. Rather, the consumer should be informed of the risks, as well as the benefits, so that he or she can decide for themselves.

In addition, there are a number of other problems associated with a simple risk–benefit analysis. An important aspect of this is that risk–benefit analyses frequently do not take into account the distribution of risks and benefits. Thus, it may not be appropriate to go ahead with a project, despite the fact that the overall benefits outweigh the risks, if the benefits are all mine and the risks all yours. A further point is that risk–benefit analysis usually ignores the complexity of people's attitudes to risk (and if it doesn't simply ignore these complexities it often struggles to do justice to them). For example, people's attitudes to voluntary risk tend to differ significantly from their attitudes to involuntary risk. They are much more willing to accept voluntary risks compared to involuntary risks. Chauncey Starr[8] writes: "we are loathe to let others do unto us what we happily do to ourselves". Finally, problems arise when comparing different types of benefits and risks. For example, how do we compare convenience, pleasure, or financial benefits with risks to life and limb? This isn't to say we can't make these comparisons – we actually make them frequently. But we make them on a case-by-case basis, and we make them for ourselves. Few people would be willing to let others make these judgements on our behalf, and furthermore we would be suspicious of any claims that these judgements could be reduced to a simple algorithm.

It is because of these various flaws that we are likely to turn to alternatives to risk–benefit analysis, such as the 'informed consent' model.

Risk–benefit analysis – the 'informed consent' model

If we are considering a case in which the use of a particular product involves risks only for the person using it, then the 'informed consent' model can be applied straightforwardly. Those supplying the product have an obligation to inform the consumer of the risks and the consumer can then make an informed decision. However, if we consider the public's concerns regarding nanotechnology or GM crops, then it is less clear how the informed consent model should be applied. This is because it is not only the consumers of the products of these technologies who may be put at risk.

We must decide on an appropriate response to risk in situations of this nature where a very large number of people could be affected. One response would be to state that it is wrong to impose risk on any non-consenting individuals, and therefore we would need consent from *everyone*. However, this seems to set an unreasonably high standard. Furthermore, we could ask: why should the burden of obtaining consent be on those in favour of change, rather than on those who wish to maintain the status quo? Whittaker[9] cites evidence that life expectancy in Canada increased from 60 years in 1931 to 69.3 years in 1971: "It is postulated that this increase in life expectancy is due to general progress, and not just to progress in one specific area". If we accept this, we might argue that Luddites averse to change associated with perceived risks could be imposing the risk of shorter life expectancy on others by refusing to consent to the introduction of new technologies. Furthermore, there will be situations where any of the options available to the general populace will involve risk: under these circumstances a decision procedure that does not require unanimity will be required.

With regard to risk imposed on large numbers of people, the analogy with the 'informed consent' model in medical ethics is not straightforward. But this needn't lead to the conclusion that the analogy with medical ethics is useless in these cases. The important point is that it is simply not acceptable for 'experts' to impose on others their views of what is best, without taking note of the concerns or wishes of those who will be affected.

From this brief discussion it should be clear that there are good reasons for scientists to respect public opinion and decisions reached democratically, even when they disagree with these decisions or opinions.

10.5 – Ethical choices in your business venture

As stated at the beginning of this chapter, when you are developing your ideas at the start of a business venture you have a great opportunity to do something positive, and to be a pioneer in ethics as well as in enterprise. Furthermore, if you manage to develop an idea that is both ethical and commercial, the ethical and the commercial aspects needn't be in conflict with each other. The ethical dimension of the company or of the product may actually be a selling point. Consider the examples of Fairtrade and Ecover. Many consumers want to support efforts to guarantee a fair wage to workers in developing countries and they want to help protect the environment where they can. They therefore choose to buy some products for ethical reasons. People do not, typically, buy Ecover laundry liquid because it gets clothes cleaner than other brands; they buy it because of Ecover's environmental policy (www.ecover.com/gb/en/About/). Similarly, people don't typically buy Fairtrade bananas because they taste better than other bananas; they buy them because of Fairtrade's commitment to pay the farmers a fair price for their products (www.fairtrade.org.uk).

Even if you don't develop an idea that has ethics as one of its main selling points, you will probably want to avoid becoming involved with something that you consider to be *un*ethical. This final section discusses a line of argument that people often use to persuade themselves to do something that they would otherwise avoid: 'if I don't do it, someone else will' (see *Case study 10.3*).

Case study 10.3: Biological weapons development

Lowell has just obtained a PhD in Microbiology and is now trying to decide on a career. During the latter stages of his PhD studies he realised that the results of his research could be used in the development of biological weapons. Lowell is not quite sure of his own views on biological weapons but he knows that a lot of people would criticise him if he decided to work in this area. He has been offered jobs at other firms, and he has another idea for a project, but he suspects that he could make most money by working on his idea for a biological weapon. However, he is not sure if he can justify this decision ethically.

Eventually he concludes that, while his idea provides an original insight, it is not *that* original and it seems likely that, sooner or later, someone else will have the same insight. The person sharing Lowell's insight may not be concerned about the ethical issues and the weapons may therefore be developed anyway. In short, Lowell concludes 'if I don't do it, someone else will', and he decides to go ahead with the project.

In *Case study 10.3,* the first thing to note about Lowell's assertion that 'if I don't do it someone else will' is that it is probably true. If we accept that Lowell's insight was not particularly original, and that we can reasonably expect others to have similar ideas, then it would be naïve to think that, provided Lowell doesn't do the work, this particular form of biological weapon will not be produced. However, the interesting question is not whether the statement is true but rather: can the statement justify Lowell's decision? Although it is stipulated here that the statement is true, it doesn't seem to justify Lowell's decision – at least, many people will not be satisfied by his reasoning and decision to proceed with the project. Indeed, if Lowell was a real person, we might even question his sincerity, and ask if he has any faith in the argument himself. We might suspect that his argument is nothing more than sophistry and that he has made his decision based on the economic benefits to himself.

Of course, this is all rather unfair. Lowell is a fictional character, so I could invent any story for him. If this was a real case, for all I know, Lowell could be fully and sincerely committed to the argument he has put forward.

My point is, simply, that many people are very suspicious of this type of argument, and rightly so. After all, the same argument could be used to justify all sorts of irresponsible behaviour or even atrocities.

Suppose, for example, that Lowell found himself in a situation where he was ordered to execute an innocent man. We can assume, for the sake of argument that if Lowell didn't do it someone else would. Lowell's refusing to shoot the man will therefore not change anything as the man is going to die anyway. If we accept the argument that Lowell originally used to justify his business decision, it shouldn't matter whether Lowell refuses to shoot the man or not. He should be entirely indifferent. He could just shrug his shoulders and think, why not? I might as well. Indeed, we can suppose that, actually, Lowell had always wanted to fire a gun. Admittedly, he had always imagined that he would just shoot at a target in a firing range, or at a can on a log, rather than at another person. Nevertheless, the fact is he had always wanted to fire a gun, just to see what it felt like, and now he has the opportunity – and the man is going to die anyway, so why not? On this account, he should feel no pangs of guilt or remorse because he has done nothing wrong.

But this reasoning seems absurd. The consequences of an action are clearly a relevant consideration, but they are not the only relevant moral consideration. If ethics were just a matter of weighing up the consequences, then we might plausibly conclude that Lowell should start his career by developing biological weapons. But, likewise, this would also commit us to the view that, in the scenario described above, Lowell might as well shoot the man. It makes no difference either way. Once we accept that these are the inevitable conclusions that follow from this form of reasoning we realise that ethics is not just a matter of weighing up the consequences. Therefore, merely stating that 'if I don't do it, someone else will', will not be sufficient to justify otherwise dubious decisions.

Of course, in a single chapter, it is not possible to cover all of the ethical issues that relate to bioscience and entrepreneurship. Indeed, there are a number of big issues in the biosciences that I have not attempted to address such as the use of animals and embryos in research, gene cloning and genetically modified foods, but several suggestions for further reading in these areas are provided in *Section 10.7* – Additional resources.

10.6 – References

1. **Schaper M** (2005) *Making Ecopreneurs: Developing Sustainable Enterprise.* Ashgate Publishing, Aldershot.
2. **Crane A and Matten D** (2004) *Business Ethics.* Oxford University Press, Oxford.
3. **Hardin G** (2000) The tragedy of the commons. In: *Environmental Ethics: An Introduction with Readings,* Benson J (ed.) Routledge, London.

4. **Boseley S** (2006) Drugs firm blocks cheap blindness cure: company will only seek licence for medicine that costs 100 times more. *The Guardian*, 17 June 2006 (www.guardian.co.uk/medicine/story/0,,1799772,00.html)
5. **Chaio J** (1993) Intellectual property: a tenuous concept. *Science*, **262** (5132): 366.
6. **Isaak R** (2005) The making of the ecopreneur. In: *Making Ecopreneurs,* Schaper M (ed.). Ashgate Publishing, Aldershot.
7. **Popper K** (2000) *The Lesson of This Century*. Routledge, London.
8. **Starr C** (1969) Social benefits versus technological risks. *Science*, **165** (3899): 1235.
9. **Whittaker JD** (1986) Evaluation of acceptable risk. *The Journal of the Operational Research Society,* **37**: 545.

10.7 — Additional resources

Animal research

Singer P (1993) *Practical Ethics, second edition*. Cambridge University Press, Cambridge. Singer's arguments regarding animal ethics have been very influential and very controversial. Chapters 3 and 5 provide an accessible summary of the arguments originally presented in his *Animal Liberation*. Essentially, Singer argues that we are 'speciesist' if we try to justify our treatment of animals merely by appealing to the fact that they are not human. Singer argues that we cannot justify our treatment in any other way — unless we are also willing to treat some humans the same way.

Hills A (2005) *Do Animals Have Rights?* Icon Books, Cambridge. A more moderate discussion of animal ethics, especially chapter 13.

Business ethics (including corporate social responsibility)

Crane A and Matten D (2004) *Business Ethics.* Oxford University Press, Oxford. An accessible introduction to business ethics, covering big topics like sustainability and globalisation, as well as more everyday issues like fair wages and employee privacy.

Schaper M (2005) *Making Ecopreneurs.* Ashgate Publishing, Aldershot. A collection of articles exploring the idea that being environmentally friendly can give firms a competitive advantage.

Environmental ethics

Benson J (2000) *Environmental Ethics: An Introduction with Readings.* Routledge, London. An accessible introduction to environmental ethics that includes introductions to a range of topics in environmental ethics by Benson, as well as a collection of seminal papers from other authors.

Light A and Rolston III H (2003) *Environmental Ethics: An Anthology.* Blackwell Publishing, Oxford. A large collection of seminal papers, covering many areas of environmental ethics.

Singer P (1993) *Practical Ethics, second edition*. Cambridge University Press, Cambridge. Chapter 10 provides a short introduction to environmental ethics, focusing in particular on the question of *why* we should care about the environment: to protect our own self-interest, for the benefit of future generations (of humans), for the sake of all sentient beings, or for the sake of the planet itself?

Genetics

Burley J (1999) *The Genetic Revolution and Human Rights: The Oxford Amnesty Lectures 1998.* Oxford University Press, Oxford. A collection of papers by leading philosophers addressing genetic screening, enhancement and cloning, etc.

Meek J (2000) Patenting our genes. *The Guardian*, June 26, 2000 (www.guardian.co.uk/genes/article/0,2763,336605,00.html).

Meek J (2000) Patenting our genes. *The Guardian*, November 15 2000 (www.guardian.co.uk/genes/article/0,2763,397385,00.html)

Nuffield Council on Bioethics, *Genetically Modified Crops: The Ethical and Social Issues.* (www.nuffieldbioethics.org/). A thorough discussion of the ethical issues relating to GM crops written by academics from a range of subjects, addressing not only the safety issues, but also issues relating to intellectual property, the impact on developing countries, and the responsibilities of scientists when advising policy makers.

The Guardian Special report: the GM debate (www.guardian.co.uk/gmdebate/0,,178400,00.html). This link contains an archive of the various articles that have been published in *The Guardian* on the topic of GM crops.

Human research

De Castro LD (2001) Ethical issues in human experimentation. In: *Companion to Bioethics*, Kuhse H and Singer P (eds). Blackwell Publishing, Oxford. This paper gives a simple introduction to some of the issues relating to research on humans, such as consent, confidentiality and exploitation.

Hope T (2004) *Medical Ethics: A Very Short Introduction*. Oxford University Press, Oxford. In chapter 8 of this book, Hope considers the ethical issues relating to research on humans in developing countries, using placebos as the control, arguing that this practice can be justified.

Lurie P and Wolfe SM (2006) Unethical trials of interventions to reduce perinatal transmission of the human immunodeficiency virus in developing countries. In: *Bioethics: An Anthology*, Kuhse H and Singer P (eds). Blackwell Publishing, Oxford. This paper considers the ethical issues relating to research on humans in developing countries, using placebos as the control, arguing that it is unethical.

Intellectual property

Hettinger EC (1989) Justifying intellectual property. *Philosophy and Public Affairs*, **18:** 31–52. A very good discussion of intellectual property. More detailed and more sophisticated (and therefore more difficult) than the discussion in this chapter. But Hettinger, in my opinion, is too quick to reject the rights-based approach.

Sharp Paine L (1991) Trade secrets and the justification of intellectual property: a comment on Hettinger. *Philosophy and Public Affairs*, **20:** 247–263. An interesting response to Hettinger that does a good job of explaining where Hettinger went wrong regarding rights.

Nanotechnology

Hunt G and Mehta M (2006) *Nanotechnology: Risk, Ethics and Law*. Earthscan Publications Ltd, London. This collection of papers includes a number which discuss some of the ethical issues involved in nanotechnology, including risk, public understanding and patents.

www.guardian.co.uk/life/nanojury/0,16014,1483441,00.html

The Guardian nanojury site contains an archive of the various articles that have been published in *The Guardian* on the topic of nanotechnology.

Responsibilities of the scientist

Kitcher P (2003) *Science, Truth and Democracy*. Oxford University Press, Oxford. In chapter 7, Kitcher argues against what he calls the "myth of purity" – the idea that scientists just do science, and are concerned only with discovering truth, and that they are not required to think about politics or ethics. In chapter 14, he considers the special obligations of scientists given that we live in an imperfect world.

Risks, benefits and public opinion

Ethics and Risk Assessment. *Perspectives on the Professions*, **9** (2): January 1990 (www.iit.edu/departments/csep/perspective/pers9_2jan90.html). This special edition of the journal contains a number of short and accessible papers discussing the limits of risk–benefit analysis.

Kitcher P (2003) *Science, Truth and Democracy.* Oxford University Press, Oxford. Chapter 10 of this book presents a very interesting approach to science, defending an approach that would give the public a much more significant role in the direction of science. In chapter 14, Kitcher concedes that we don't live in a perfect world, and considers the responsibility of scientists working in an imperfect world (working, for example, on a project that – while generally beneficial – may be harmful to some).

Schinzinger R and Martin MW (2000) *Introduction to Engineering Ethics.* McGraw-Hill Higher Education, Harlow. Chapters 3 and 4, although discussing engineering rather than the biosciences, address many of the same ethical issues.

Wolff J (1996) *Introduction to Political Philosophy.* Oxford University Press, Oxford. Chapter 3 covers the value of democratic decision-making, even when such decisions conflict with expert opinion.

General bioethics anthologies (covering many of the topics above)

Kuhse H and Singer P (2001) *A Companion to Bioethics.* Blackwell Publishing, Oxford.

Kuhse H and Singer P (2006) *Bioethics: An Anthology.* Blackwell Publishing, Oxford.

10.8 – Acknowledgements

David Adams, Caroline Cobden, Graeme Gooday, Joanne Koukis and George Koukis.

Learning outcomes

Key learning points from this chapter are:

- There are numerous ethical issues that are relevant both to the scientist and to the entrepreneur.

- A proper consideration of ethics involves the use of a range of analytical skills and careful consideration of the various arguments that may be deployed for any given situation; it is much more than a simple, introspective search of one's conscience.

- You should always undertake extensive research to ensure that you are fully aware of the differing sides of arguments before you debate ethical issues. The material provided in the 'Additional resources' section should help inform future discussions.

Index